Nora Bandi

miRNA involvement in cell cycle control of non-small cell lung cancer

Nora Bandi

miRNA involvement in cell cycle control of non-small cell lung cancer

Südwestdeutscher Verlag für Hochschulschriften

Impressum/Imprint (nur für Deutschland/only for Germany)
Bibliografische Information der Deutschen Nationalbibliothek: Die Deutsche Nationalbibliothek verzeichnet diese Publikation in der Deutschen Nationalbibliografie; detaillierte bibliografische Daten sind im Internet über http://dnb.d-nb.de abrufbar.

Alle in diesem Buch genannten Marken und Produktnamen unterliegen warenzeichen-, marken- oder patentrechtlichem Schutz bzw. sind Warenzeichen oder eingetragene Warenzeichen der jeweiligen Inhaber. Die Wiedergabe von Marken, Produktnamen, Gebrauchsnamen, Handelsnamen, Warenbezeichnungen u.s.w. in diesem Werk berechtigt auch ohne besondere Kennzeichnung nicht zu der Annahme, dass solche Namen im Sinne der Warenzeichen- und Markenschutzgesetzgebung als frei zu betrachten wären und daher von jedermann benutzt werden dürften.

Verlag: Südwestdeutscher Verlag für Hochschulschriften GmbH & Co. KG
Heinrich-Böcking-Str. 6-8, 66121 Saarbrücken, Deutschland
Telefon +49 681 37 20 271-1, Telefax +49 681 37 20 271-0
Email: info@svh-verlag.de

Approved by: Bern, Graduate School for Cellular and Biomedical Sciences of the University of Bern, PhD thesis, 2010

Herstellung in Deutschland:
Schaltungsdienst Lange o.H.G., Berlin
Books on Demand GmbH, Norderstedt
Reha GmbH, Saarbrücken
Amazon Distribution GmbH, Leipzig
ISBN: 978-3-8381-2979-2

Imprint (only for USA, GB)
Bibliographic information published by the Deutsche Nationalbibliothek: The Deutsche Nationalbibliothek lists this publication in the Deutsche Nationalbibliografie; detailed bibliographic data are available in the Internet at http://dnb.d-nb.de.

Any brand names and product names mentioned in this book are subject to trademark, brand or patent protection and are trademarks or registered trademarks of their respective holders. The use of brand names, product names, common names, trade names, product descriptions etc. even without a particular marking in this works is in no way to be construed to mean that such names may be regarded as unrestricted in respect of trademark and brand protection legislation and could thus be used by anyone.

Publisher: Südwestdeutscher Verlag für Hochschulschriften GmbH & Co. KG
Heinrich-Böcking-Str. 6-8, 66121 Saarbrücken, Germany
Phone +49 681 37 20 271-1, Fax +49 681 37 20 271-0
Email: info@svh-verlag.de

Printed in the U.S.A.
Printed in the U.K. by (see last page)
ISBN: 978-3-8381-2979-2

Copyright © 2011 by the author and Südwestdeutscher Verlag für Hochschulschriften GmbH & Co. KG and licensors
All rights reserved. Saarbrücken 2011

Content

1. **SUMMARY** 3
2. **INTRODUCTION** 5
   2.1. PATHOPHYSIOLOGY OF LUNG CANCER 5
      2.1.1. Epidemiology and incidence statistics of lung cancer 5
      2.1.2. Classification of lung cancer 5
      2.1.3. Pathogenesis of lung cancer 6
      2.1.4. Current concepts in the treatment of lung cancer 6
   2.2. MICRORNA'S 8
      2.2.1. Definition, overview and nomenclature 8
      2.2.2. Biogenesis and function 9
      2.2.3. Expression and target recognition 11
      2.2.4. miRNAs and cancer 13
      2.2.5. miRNA involvement in the tumourigenesis of lung cancer 15
      2.2.6. miRNA involvement in cell cycle regulation 19
      2.2.7. miR-15a/16 21
      2.2.8. miR-34 23
      2.2.9. miRNAs as diagnostic markers and therapeutics 24
3. **AIM OF THE PROJECT** 26
4. **RESULTS** 29

   MANUSCRIPT 1: miR-15a and miR-16 are implicated in cell cycle regulation in a Rb-dependent manner and are frequently deleted or down-regulated in non-small cell lung cancer 29

   MANUSCRIPT 2: miR-34a and miR-15a/16 are co-regulated in non-small cell lung cancer and control cell cycle progression in a synergistic and Rb-dependent manner 53

   MANUSCRIPT 3: Linking microRNAs to aberrant cell cycle control in cancer 77

5. **DISCUSSION** 95
6. **OUTLOOK** 100
7. **REFERENCES** 102
8. **ACKNOWLEDGMENTS** 111

1. Summary

Lung cancer is the leading cause of cancer associated death worldwide and is primarily caused by tobacco smoke. Non-small cell lung carcinoma (NSCLC) accounts for 80% of lung cancer and is further subdivided into two major types, squamous cell carcionoma and adenocarcinoma. These tumours are characterised by multiple and heterogeneous genetic and epigenetic alterations of genes involved in cell cycle control and/or apoptosis. Despite improvements in the therapy of NSCLC, the overall five-year survival rate remains less than 15%. Therefore new therapeutic strategies are required to improve clinical outcome. The fact that microRNAs (miRNAs) are often deregulated in many diseases including cancer opens a new field of potentially powerful novel therapeutic approaches.

MicroRNAs are negative regulators of gene expression at the post-transcriptional level that are frequently involved in carcinogenesis. Although many miRNAs form part of integrated networks, little information is available how they interact with each other to control biological processes. To address this question the role of *miR-15a/16* and *miR-34a* in the tumourigenesis of non-small lung cancer was investigated in this thesis.

We show that *miR-15a/16* and *miR-34a* are co-regulated in the majority of squamous cell carcinomas and adenocarcinomas of the lung. In most cases, both miRNAs are significantly down-regulated, indicating that they play an important role as tumour suppressors. Co-regulation is not directly linked nor is it due to defects in miRNA processing. In NSCLC, expression of *miR-15a/16* inversely correlates with the expression of cyclin D1. Important G_1 cyclins like cyclin D1, D2 and E1 are directly regulated by physiological concentrations of *miR-15a/16* in NSCLC cell lines. Recent publications demonstrated that *miR-34a* is involved in the regulation of a similar set of target genes, including G_1 cyclins. Consistent with these findings, over-expression of *miR-15a/16* and *miR-34a* induces cell cycle arrest in G_1/G_0. Cells lacking Rb are resistant to *miR-15a/16*- or *miR-34a*-induced cell cycle arrest, while re-introduction of functional Rb resensitizes these cells to miRNA activity. Thus, cell cycle arrest induced by *miR-15a/16* or *miR-34a* depends on the expression of Rb, confirming that G_1 cyclins are major targets of *miR-15a/16* and *miR-34a* in NSCLC.

Summary

Currently, research is focused primarily on identifying novel targets of individual miRNAs, while the analysis of combinatorial effects of deregulated miRNAs in specific cancer types is still not uncovered. *miR-15a/16* and *miR-34a* are functionally related. Despite the fact that they contain distinct seed sequences, they share common targets and are implicated in similar processes including cell cycle control. Thus, using the example of *miR-15a/16* and the closely related *miR-34a*, we investigated how miRNAs with overlapping functions affect biological processes in a combinatorial mode. We demonstrate that *miR-15a/16* and *miR-34a* act synergistically in inducing cell cycle arrest. Both miRNAs act on common targets in an additive rather than synergistic manner, while the synergistic effect of these miRNAs on cell cycle arrest is due to down-regulation of targets unique to either one of these miRNAs. This is based on the finding that knocking-down *cyclin E1*, which is a unique target of *miR-15a/16*, by RNA interference, completely abrogated the synergistic effect exerted by the combined action of both miRNAs. Thus miRNAs act in a synergistic manner since in combination they down-regulate more targets than each miRNA alone.

In conclusion our results indicate that *miR-15a/16* and *miR-34a* are synergistically implicated in cell cycle control and likely contribute collectively to the tumourigenesis of NSCLC. We propose two alternative pathways by which NSCLC cells escape *miR-15a/16*- and *miR-34a*-induced cell cycle arrest: (i) co-repression of *miR-15a/16* and *miR-34a* or (ii) inactivation of the *Rb* gene.

In addition our results suggest that the combination of miRNAs, which form part of the same network, should be considered for assessing a biological response rather than individual miRNAs. Since *miR-34a* and *miR-15/16* are frequently down-regulated in the same tumour tissue, these findings may have important therapeutic implications for microRNA-based treatment of Rb-proficient NSCLC.

2. Introduction

2.1. Pathophysiology of lung cancer

2.1.1. Epidemiology and incidence statistics of lung cancer

Lung cancer is the most common cancer in humans and the leading cause of cancer death worldwide with 1.61 million new cases and 1.38 million fatalities per year (http://info.cancerresearchuk.org/cancerstats/world). Despite major geographic, racial and gender differences in the incidence of lung cancer, most cases are caused by tobacco smoke.[1]. Not all cases of lung cancer are associated with active smoking, but the role of passive smoking is increasingly being recognised as a risk factor for lung cancer. The occurrence of lung cancer in non-smokers, who account for 15% of cases, is often attributed to a combination of genetic factors and environmental factors including exposure to radon gas, asbestos, and air pollution. While the prevalence of smoking is decreasing in the USA and Europe, it is raising in developing countries, where an increased incidence of lung cancer is expected in the next few years. Despite improvements in therapy ~ 80-90% of lung cancer patients will die from their disease.

2.1.2. Classification of lung cancer

Lung cancer includes two major types, small cell lung cancer (SCLC) and non-small cell lung cancer (NSCLC), which can be distinguished by histology. This classification has important implications for clinical management and prognosis of the disease.

SCLC is the less common form of the two major histological types (15-20%). It usually starts in the larger breathing tubes and originates from neuroendocrine cells.

NSCLC accounts for ~ 80% of all lung cancer cases [2] and can be further divided into three major subgroups, adenocarcinoma, squamous cell carcinoma and large cell carcinoma of the lung [3]. Squamous cell carcinoma usually arises from the major bronchi. Since there is no squamous epithelium in the normal lung, these tumours originate from metaplastic changes of lung epithelial cells. Adenocarcinomas and large cell carcinomas are peripherally located and arise from distant airway bronchioles and alveoli. Adenocarcinoma, the most common subtype of NSCLC [3], originates from progenitor cells of the bronchioles or alveoli. Large cell carcinoma represents a heterogeneous group of poorly differentiated tumours.

2.1.3. Pathogenesis of lung cancer

Cancer cells develop mechanisms to grow, invade and metastasize. Lung cancer is characterized by multiple and heterogeneous genetic and epigenetic alterations. Similar to many other cancers, lung cancer is initiated by the activation of oncogenes or/and the inactivation of tumour suppressor genes involved in various processes including cell cycle control, DNA repair and apoptosis. Common oncogenes that contribute to the development of lung cancer include *c-Myc, RAS, EGFR, cyclin D1* and *Bcl-2* [4]. Tumour suppressor genes that are frequently inactivated include *p53, Rb* and *p16INK4a*. Many of the molecular changes are common in NSCLC and SCLC, while others predominate in one form or the other (reviewed by ref [5]). For example *p53* is frequently inactivated in both forms, while *k-RAS* is mutated only in NSCLC and never in SCLC. Recently it became clear that alterations of microRNA expression may also contribute to the tumourigenesis of lung cancer [6] (see chapter 2.2.5).

Altered expression of oncogenes or/and tumour suppressor genes in neoplastic lung tissues may be due to genetic and epigenetic alterations or deregulations of transcription factors, but the relevant molecular mechanisms driving the aggressive biological behaviour of these tumours are largely unknown.

2.1.4. Current concepts in the treatment of lung cancer

Despite advances in the diagnosis and therapy of lung cancer, the prognosis for patients has only improved marginally in the past few years and the overall 5-year survival rate remains less than 15% [7]. Explanations for the poor survival include late presentation of the disease, a lack of markers for early detection and both phenotypic and genotypic heterogeneity between patients and between tumour regions of the same patient.

Lung cancer may be detected by chest radiography and computed tomography (CT scan) and is usually confirmed by histology from a biopsy. The majority of lung cancer cases are in an advanced stage at diagnosis. Treatment and prognosis depend on the histological type of cancer, the tumour stage and the patient's performance status. Small cell lung carcinomas, which tend to metastasize early, are treated primarily with chemotherapy and radiation therapy, as surgery has no demonstrable influence on survival. Surgery is generally regarded as the best treatment option in NSCLC, but unfortunately only a minority of patients are amenable to potentially curative surgical intervention [8]. Usually surgery is only an option in early stage NSCLC limited to one half of the lung. Patients with good performance status with unresectable, locally advanced NSCLC, undergo radical thoracic radiotherapy or combined chemo-radiotherapy. However such strategies do only marginally prolong survival. The majority of patients, essentially those with metastatic disease, are managed with palliative therapy regimes based primarily on chemotherapy [9]. Recent advances in the understanding of the molecular pathology of lung cancer led to the development of new therapeutic strategies targeted against molecular mechanisms that mediate tumour growth. Among those, tyrosine kinase inhibitors specific for the epidermal growth factor receptor or the vascular endothelial growth factor receptor have provided some clinical benefits. However, although various new chemotherapeutic agents were developed in the past decades, currently available chemotherapy for advanced lung cancer still has only a limited benefit for the survival of patients [10] and the emergence of resistance to chemotherapy remains a major problem in the treatment of patients with lung cancer. Therefore new innovative therapeutic strategies that are more effective than older chemotherapeutic drugs are required to improve clinical outcome.

Altogether, current treatment regimes are often limited to the improvement of the quality of life of lung cancer patients while the primary prevention of lung cancer remains the prevention of smoking initiation.

2.2. microRNA's

microRNAs (miRNAs) play important roles in human carcinogenesis. The fact that miRNAs are frequently deregulated in tumour cells opens a new field of potentially powerful novel therapeutic approaches for the treatment of human cancer.

2.2.1. Definition, overview and nomenclature

The relatively recent discovery of several types of non-protein-coding RNAs, such as small nucleolar RNAs, small interfering RNAs and miRNAs added an additional layer of complexity to the understanding of gene regulation in multicellular eukaryotes.

miRNAs are highly conserved small noncoding RNA molecules of 19 to 24 nucleotides that negatively regulate gene expression at the posttranscriptional level. This novel class of RNAs was first discovered in the roundworm *Caenorhabditis elegans* in 1993 [11]. In the following years miRNAs have been found in nearly every eucaryotic system examined. miRNAs have been identified in humans in the year 2000/2001 [12, 13].

Currently 940 human miRNA sequences are registered in the miRBase database (http://microrna.sanger.ac.uk/sequences/) and it is estimated that the human genome harbours about 1000 of these sequences (i.e. 1-5% of all predicted genes in the genome) [14].

MicroRNAs play an important regulatory role in numerous biological processes as diverse as early development [15], cell proliferation [16], cell death [17], cell metabolism [18], immune response [19], viral replication [20] and cell differentiation [21]. Many miRNAs are evolutionarily conserved across multiple species implying the importance of these molecules as modulators of critical biological pathways [22, 23]. It even seems that the development of multicellular organisms correlates with the appearance of miRNAs [24].

Although the first publication of a miRNA appeared seventeen years ago [11], the knowledge about the complexity and diversity of this class of small regulatory RNAs has exploded in the past few years and the number of publications dealing with this subject increased almost exponentially every year. Research has mainly been focused on the understanding how, when, and where miRNAs are produced and function in cells, tissues, and organisms. However, their precise biological relevance is still largely unknown.

MicroRNA names have a prefix indicative of the species and a unique numerical suffix. For example, *hsa-miR-21* refers to the mature *miRNA-21* of humans, whereas *mmu-miR-21* refers to the corresponding murine ortholog. Mature miRNAs whose sequences differ at only few nucleotide positions are thought to regulate the same target genes. These miRNAs are assigned the same numerical suffix along with a distinguishing alphabetical identity. For example hsa-*miR-15a* and hsa-*miR-15b* are the products of different miRNA genes differing from each other at only four nucleotide positions. Mature miRNAs that have an identical sequence but arise from different genes are given additional numerical suffixes (e.g. hsa-*miR-16-1* and hsa-*miR-16-2*).

miRNA cloning studies sometimes identify two distinct sequences, which originate from the same predicted precursor. When one miRNA is clearly predominantly expressed, then this is indicated with names of the form *miR-34b* (the predominant sequence) and *miR-34b** (the less abundant form from the opposite side of the precursor).

When it is not clear, which sequence is the predominant one, names like *miR-125-5p* (from the 5'arm of the precursor) and *miR-125-3p* (from the 3' arm of the precursor) are given (http://www.mirbase.org/help/nomenclature.shtml).

For historical reasons the first discovered miRNAs *let-7* and *lin-4* are obvious exceptions to this nomenclature.

2.2.2. Biogenesis and function

In general miRNA genes are transcribed by RNA polymerase II and therefore contain a 5' cap and a 3' poly(A) tail [25, 26]. However, some miRNAs, located in Alu sequences or other repetitive elements, are transcribed by RNA polymerase III [27]. The generation of active single-stranded miRNAs from long partially double stranded transcripts is a multi-step process (figure 1).

The initial, long primary microRNA transcripts (pri-microRNAs) form typical fold-back structures, which are processed into ~70 nucleotide precursor miRNAs (pre-miRNAs) by the microprocessor complex consisting of the RNase III enzyme Drosha and the dsRNA-binding protein DGCR8 [28]. Some miRNAs, so-called miRtrons, are initially processed independent of Drosha and DGCR8. These miRNAs are located in introns of other genes, which is the reason why their initial processing is achieved by splicing factors [29].

The generated pre-miRNAs are exported from the nucleus into the cytoplasm by the Ran-GTP dependent transporter Exportin-5 [30]. In the cytoplasm, pre-miRNAs are subjected to a further cleavage by the RNase III enzyme Dicer, giving rise to a double-stranded RNA of 19-24 nucleotides with a 1–4 nt 3' overhang at either end [31]. By analogy to Drosha, cofactors such as TRBP and PACT are necessary for Dicer activity [32].

Next, the generated miRNA duplex is incorporated into the RNA induced silencing complex (RISC), a multiprotein complex that separates the mature strand from the passenger strand and facilitates the binding of the miRNA to mRNAs at sequences that are complementary to the mature miRNA.

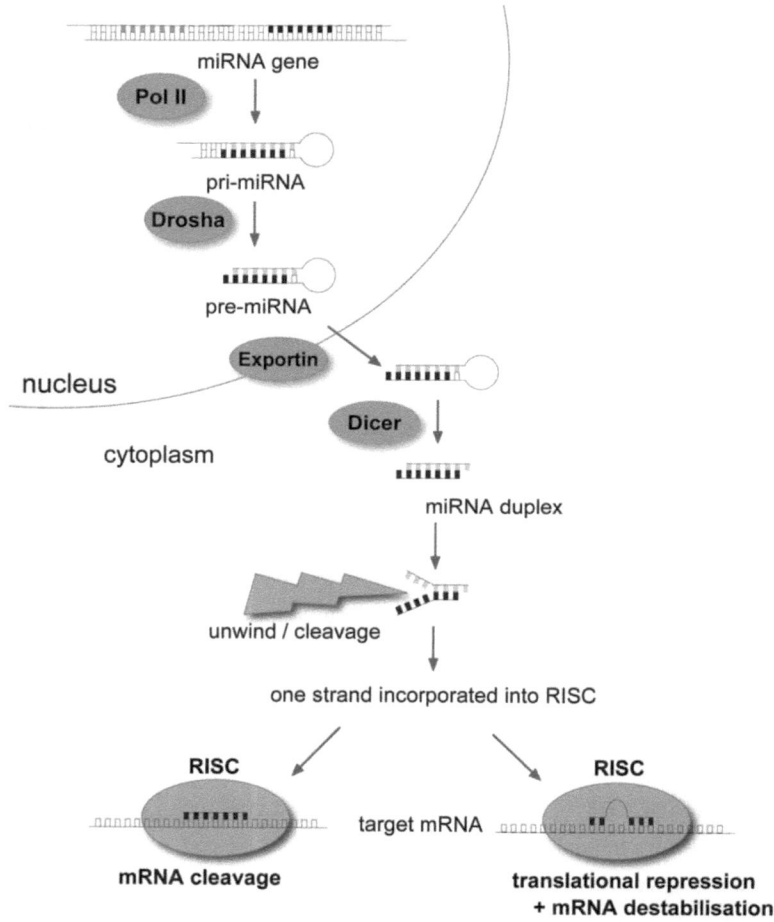

figure 1. biogenesis and functions of miRNAs

The selection of the mature strand from the miRNA duplex appears to be based primarily on the stability of the termini of the two ends of the miRNA duplex. The strand with the lower stability preferentially associates with RISC and thus becomes the active miRNA, while in the majority of cases the other strand is degraded [33]. Mature miRNAs can undergo further modifications such as 3' uridylation or adenylation and nucleotide substitutions [34], which can affect their stability as well as their function. Once loaded into the RISC, mature miRNAs pair with mRNAs to direct post-transcriptional repression. miRNAs that bind with perfect or nearly perfect complementarity to protein coding mRNAs induce the RNA-mediated interference (RNAi) pathway [35]. Briefly, mRNA transcripts are cleaved by ribonucleases in the miRNA-associated RISC, which results in the degradation of target mRNAs. This mechanism is particularly found in plants but has also been shown to occur rarely in mammals [36, 37]. More commonly in mammals though, miRNAs direct translational repression, mRNA destabilization or a combination of the two. miRNAs involved in this second mechanism form imperfect base-pairing to a target mRNA region, mostly located in the 3' UTR. However recent studies suggest, that in some cases, binding to 5' UTRs or coding regions can occur as well [38]. Although significant progress has been made in understanding the biological function of miRNAs, the mechanistic details by which miRNAs repress translation is still unclear. Plenty of studies have been published in recent years, which propose various molecular models of how miRNAs impede translation. However, the fundamental question whether repression occurs at translation initiation or post initiation has not yet been resolved. Currently it is suggested that regulation occurs at various steps along the translational process (reviewed in refs [39, 40]). Early studies of animal miRNAs indicated that translational repression is not accompanied by mRNA destabilization. However recent findings indicate that miRNA-dependent regulation of mRNA stability due to enhanced deadenylation, decapping, and exonucleolytic digestion may be more common than previously supposed. In fact this mode of gene regulation could even be the predominant reason for reduced protein output [41]. It is still unclear if enhanced mRNA degradation and translational repression are mutually exclusive or coupled processes. At present it is believed that the final type and level of repression is a sum of events from multiple pathways [40].

Interestingly a recent report even indicates, that miRNAs can not only repress but also increase target mRNA translation, depending on whether cells are proliferating or arrested in the cell cycle [42]. However, the precise underlying mechanism remains unknown.

2.2.3. Expression and target recognition

Most of the genome sequences encoding miRNAs are located in areas of the genome that are not associated with known genes; many are found at fragile sites of human chromosomes [43] and appear to be transcribed by their own promotor[13, 23]. Approximately one quarter of all miRNAs are encoded in introns of primary mRNA transcripts, indicating that they are co-transcribed from the promoter of the protein coding gene.

Similar to mRNAs, the expression of miRNAs is regulated by transcription factors [44, 45] and by epigenetic mechanisms like methylation of CpG islands within their promotor regions (reviewed by ref [46]). Many miRNAs are clustered and form part of the same polycistronic primary transcript, suggesting a related function [47]. For example the *miR-17-92* cluster encodes seven miRNAs the expression of which is regulated by transcription factors like c-Myc [48] and E2F1 [49]. Despite these clusters being regulated by a single promoter, not all the miRNAs contained within the same cluster are expressed at a similar level, suggesting a second level of post-transcriptional regulation [34]. This finding is in accordance with many similar reports, indicating that the abundance of mature miRNAs is a highly regulated process. Regulation occurs not only at the transcriptional level, but also at the level of processing of the primary miRNAs or precursor miRNAs [34, 50, 51].

In many cases, miRNA expression is highly tissue- and developmental-specific, consistent with a general role for miRNAs in cell differentiation [21]. Interestingly, miRNA expression profiles are similar between related tissues and distinct between unrelated tissues. For instance, heart and skeletal muscle profiles are very similar to each other, but clearly distinct from the brain miRNA profile.

As already mentioned, miRNAs do not require perfect complementarity for target recognition, which is the reason why a single miRNA can often bind to hundreds of different targets. Although the impact of one miRNA on a single mRNA is usually weak, the combinatorial repression of several different target genes may produce measurable phenotypic effects [52].

Nowadays, the expression of a large number of the predicted human miRNA genes has been confirmed, but the majority of miRNA targets remain to be identified and verified. For most miRNAs the 5' region (nucleotides 2 to 7), known as the "seed-sequence" is most critical for targeting and function [53]. Consistent with this functional importance, most mammalian miRNAs are more conserved at their 5' ends relative to their 3'ends.

In silico target prediction is mostly based on different algorithms including complementarity between the seed-sequence of the miRNA and a sequence stretch in the 3' untranslated region of the mRNA target, thermodynamics, binding site structures and sequence conservation. Because different algorithms use different restrictions, their predictions do not overlap by much, and no algorithm discovers all targets [54]. Thus, even though these algorithms are indispensable for miRNA target prediction, not all predicted targets are indeed real targets and therefore have to be confirmed experimentally. Likewise many true targets may be missed using current prediction algorithms, impeding the discovery of important targets and functions of miRNAs. The ability of a miRNA to bind to its target also depends on the concentration of either of these molecules. Thus, it has also to be shown experimentally if a miRNA is able to regulate the expression of a target at its physiological concentration.

miRNAs rarely act alone (figure 2). Moreover, many mRNA targets have seed matches for multiple miRNAs and are more effectively repressed than those containing a single miRNA binding site [37]. Thus, miRNAs not only target multiple mRNAs, but mRNAs are also targeted by multiple miRNAs. This observation coupled with the prediction that many mRNAs have target sites for many different miRNAs suggests that gene expression in various tissues and cells can be greatly influenced by the miRNA and the mRNA populations in those cells.

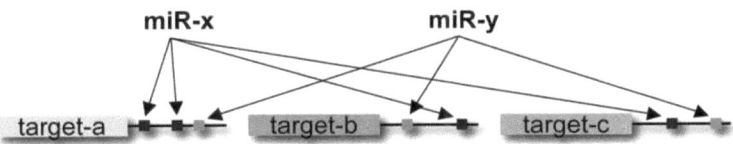

figure 2. miRNAs target multiple mRNAs and mRNAs are targeted by multiple miRNAs

2.2.4. miRNAs and cancer

A significant number of miRNA genes map to regions which are frequently altered in cancer [43] and global alterations in miRNA expression is a common event in cancer [55]. The fact that such alterations are present in virtually all tumour types, including blood-borne malignancies as well as solid tumours, suggests an important role of miRNAs in tumourigenesis. Consequently, miRNA "signatures" were discovered that distinguish between tumour and normal cells and in some cases are associated with the prognosis and the progression of cancer. A particular characteristic of such miRNA expression changes is, that the degree of dysregulation is usually very modest (in the 1.5 - 2.5 fold change range), indicating that small changes in microRNA expression can have significant effects on the phenotype. This is in accordance with the known ability of miRNAs to simultaneously influence the expression of hundreds of genes.

Aberrant regulation of miRNAs can be the consequence of genomic rearrangements, altered methylation status of their respective promotor regions or other alterations in the transcription initiation or processing of pri-miRNAs. Different oncogenes and tumour suppressor genes have been implicated in the transcription initiation or processing of miRNAs [56]. Somatic point mutations interfering with miRNA processing, miRNA stability or binding of miRNAs to target mRNAs may be additional causes for the deregulation of miRNAs in tumour tissues (reviewed by ref [57]).

The functional consequence of miRNA deregulation becomes obvious by the fact that the introduction or repression of a single miRNA can effectively contribute to tumourigenesis or tumour progression. Several functional studies using cultured cancer cells and mouse models of cancer have identified miRNAs that function as conventional tumour suppressors or oncogenes (figure 3). Oncogenic miRNAs are frequently over-expressed and contribute to tumour development by inhibiting the expression of tumour suppressor genes, whereas tumour suppressing miRNAs are frequently down-regulated, leading to the over-expression of target proto-oncogenes.

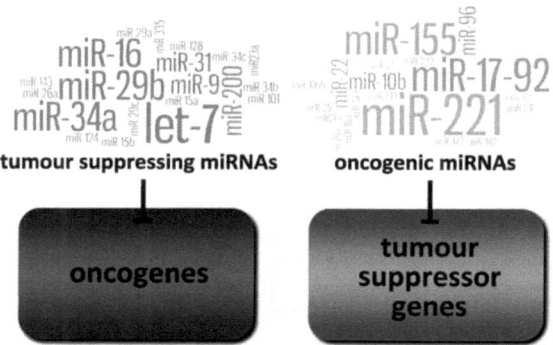

figure 3. miRNAs function as tumour suppressors and oncogenes

Examples of miRNAs with tumour–suppressing activity are miRNAs of the *let-7*, *miR-16*, and the *miR-34* families that regulate well-known oncogenes like RAS, Bcl-2 and CCND. On the other hand, oncogenic miRNAs like *miR-155, miR-21* as well as the *miR17-92* cluster regulate important tumour suppressors like PTEN, c-Maf and BIM.

Recent evidence supports the ability of miRNAs to regulate several steps during neoplastic cell transformation (reviewed in ref [58]) defined as hallmarks of cancer [59]. These include:

I) self-sufficiency of tumour cells (e.g. *miR-7, miR-125, miR-143*),
II) insensitivity to antigrowth signals (e.g. *miR-17-92, let-7, miR-34*),
III) abnormal apoptosis behaviour (e.g. *miR-21 , miR-34 , miR-127*),
IV) limitless replicative potential (e.g. *miR-372, miR-138, miR-34*),
V) induction and sustained angiogenesis (e.g. *miR-221, miR-17-92, miR-296*),
VI) invasion and metastasis (e.g. *miR-10b, miR-21, miR-221*).

In agreement with the assumption that conventional cancer genes function as such regardless of tissue of origin, the oncogenic or tumour-suppressive activity of many miRNAs (e.g. *miR-155, let-7, miR-21* and *miR-17-92*) is not limited to a particular tumour type.

However, other miRNAs have been implicated in specific types of cancers, indicating that some miRNAs may fulfil disease-specific functions.

The deregulation of some miRNAs even correlates with the tumour differentiation status, disease stage and patient outcome. This further suggests, that some miRNAs have a direct impact on tumour development.

Nevertheless, as already mentioned, a single miRNA can target hundreds of mRNAs, rendering it challenging to attribute distinct functions to specific miRNAs. Even though much is known about the aberrant expression of miRNAs in human cancer, little is known about the broad functional relevance of such aberrations.

2.2.5. miRNA involvement in the tumourigenesis of lung cancer

Lung cancer is characterised by multiple and heterogeneous genetic and epigenetic alterations. Combined Studies that examine both global changes in miRNA expression within lung tumours as well as the effects of individual miRNAs on lung cancer cell phenotype suggest that alterations in miRNA genes play a critical role in the pathophysiology of lung cancer.

Consistent with the known etiological relationship between smoking and lung cancer, smoking has been shown to alter miRNA expression [60, 61]. Consequently, such alterations seem to be involved in the deregulation of cancer-related genes involved in several processes as diverse as cell cycle control, DNA repair and apoptosis.

Currently, most miRNA-expression studies in lung cancer are focussed on NSCLC. Several studies have demonstrated the potential of miRNA profiles as diagnostic markers in NSCLC that may even help to distinguish specific subtypes, such as squamous cell carcinoma and adenocarcinoma [62, 63]. In contrast only few studies, have been focused on the role of miRNAs in the pathogenesis of SCLC [64]. This is due to the fact, that primary tissue specimens from these tumours are difficult to obtain, because SCLC normally is not surgically resected.

In high-throughput microarray-based expression analyses, several groups identified various miRNAs that are differentially expressed in NSCLC cells [65-68]. Unfortunately these studies rely on high throughput platforms that are partially limited to the lack of reproducibility and small cohort sizes. In addition, the nature of miRNA alterations identified in similar studies may often be distinct due to variations in miRNA profiling methodologies, analytical approaches, population types, histological subtypes, disease stages and grades as well as due to different methods of tissue archival and RNA extraction. Nevertheless, some of the identified miRNAs show a consistent up- or down-regulation in independent studies conducted by different groups, approving their general role in lung cancer tumourigenesis (table 1). Interestingly, a relative recent study of the changes in miRNA expression during progression from hyperplasia to invasive squamous cell carcinoma of the lung identified sixty-nine miRNAs, which are differentially expressed during the course of the disease [69]. Among these the expression of *miR-34c*, *miR-15a* and *miR-32* decreased progressively, whereas that of *miR-139* and *miR-199* was stage specific.

In addition several miRNAs have been identified which are considered as prognostic markers for overall survival or predictive markers for chemotherapy in NSCLC [70-72]. For example the levels of *let-7a*, which was the first miRNA reported to be aberrantly expressed in human lung cancer, are associated with short survival after surgery in two independent studies [65, 73]. In accordance with those findings, a single nucleotide polymorphism (SNP) in a *let-7* microRNA complementary site in the *k-RAS* 3' untranslated region has been shown to increase the risk to develop non-small cell lung cancer [74].

Recently it has been reported that miRNAs can be quantified from body fluids including serum, plasma as well as saliva and urine. In a small study of 23 cases of NSCLC patients and 17 controls, increased levels of *miR-21* were associated with cancer with sensitivity and specificity values of 70% and 100%. These values are clearly better than the values of 48% and 100% respectively, that were obtained by sputum cytology of the same patient samples [75]. Moreover, NSCLC patients can be reliably distinguished from healthy persons based on miRNA serum levels and the expression of *miR-486, miR-30d, miR-1*, and *miR-499* was significantly associated with overall survival [76]. Thus, although very preliminary, these finding implicate that miRNAs may be used to diagnose, prognosticate and monitor lung cancer via non-invasive methods.

Several of the identified deregulated miRNAs in lung cancer have been shown to target tumour suppressor genes and oncogenes (table 1) that play crucial roles in lung tumourigenesis (reviewed in refs. [6, 77]).

For example, *miR-29* is often down-regulated in lung cancer [65, 66]. *miR-29* targets DNA methyltransferases (DNMT) 3A and 3B, two key enzymes involved in DNA methylation, which are often up-regulated in several malignancies including lung cancer [113]. Some tumour suppressor genes involved in lung cancer are susceptible to epigenetic silencing by methylation [114]. Re-expression of *miR-29* in lung cancer cells restores the normal DNA methylation patterns and reduces tumour development [89]. Thus miRNAs are not only regulated by DNA methylation but also modulate DNA methylation by interfering with the DNA methylation machinery. Another example is *miR-1*, which inhibits both in vitro growth and survival of NSCLC cell lines by targeting several oncogenes including *c-Met, Pim-1, FoxP1* and *HDAC4*. Several of the deregulated miRNAs have been shown to target multiple cancer related genes (table 1) indicating that their deregulation also affects more than one process during lung cancer tumourigenesis.

Manipulation of miRNA levels has been used to control lung cancer cell survival and proliferation in vitro and in vivo [89, 115-118], supporting their crucial role in lung cancer. For example, administration of *let-7* reduces tumour formation in the lungs of animals expressing an mutated form of the *k-RAS* oncogene [119].

In summary, all these results suggest that microRNAs may play an important role in the pathogenesis and progression of NSCLC. The causes for altered expression of miRNAs in neoplastic tissues of the lung may be heterogeneous. Allelic loss is a frequent genetic alteration in lung cancer. Cytogenetic and molecular analyses have revealed deletions on chromosome arms 3p, 8p, 9p, 11p, 13q, 17p, 18q and 22q [120] where several miRNAs are located. Likewise, several promoters of tumour suppressing miRNAs are susceptible to epigenetic silencing by aberrant methylation in human tumours including lung cancer [121]. General repression of miRNAs may be due to defects in miRNA processing. Notably, reduced expression of Dicer has been detected in a fraction of NSCLC with a significant

Introduction

impact on patient survival [122]. Nevertheless, in many cases the precise molecular mechanisms of miRNA deregulation in lung cancer are unknown.

miRNA	Deregulation	Function (when expressed)	Validated targets
miR-1	down [72]	proliferation ↓ migration ↓ invasion ↓ apoptosis ↑	c-Met [72], FoxP1 [72], HDAC4 [72], Pim-1 [72], Bcl2 [78]
let-7	down [73]	proliferation ↓ apoptosis ↓	RAS [79], Myc [80], HGMA2 [81], FUSI [82]
miR-15a/16	down [83]	Rb dependent cell cycle arrest ↑	CCND1 [83-85], CCND2 [83], CCND3 [85], Bcl2 [86], CCNE1 [83, 85], CDK6 [85], WNT3A [84], Cdc25A [87], Bmi-1 [88]
miR-29	down [65, 66]	apoptosis ↑ tumourigenicity ↓ (reverts aberrant methylation)	DNMT3A-3B [89], MCL-1 [90]
miR-34a-c	down [91]	cell cycle arrest ↑ apoptosis ↑ p53 dependent stress responses ↑	CDK4/6 [92, 93], Bcl2 [94], CCND1 [92], c-Met [93], c-Myc [95], E2F3 [96], SIRT-1 [97]
miR-126	down [65]	invasion ↓ migration ↓ adhesion ↓ angiogenesis ↓	CRK [98], VEGF-A [99]
miR-128b	down [100]	EGFR-mediated proliferation ↓	EGFR [100]
miR-183	down [101]	metastasis ↓	Ezrin [101]
miR-200 family	down [102]	E-cadherin expression ↑ EMT (epithelial mesenchymal transition) ↓ metastasis ↓	ZEB1/ZEB2 [103]
miR-21	up [70]	apoptosis ↓ proliferation ↑	PTEN [104], PDC4 [105], TPM1 [106]
miR-17-92	up [107]	angiogenesis ↑ apoptosis ↓	PTEN [108], E2F1-3 [49], BIM [108]
miR-155	up [65]	proliferation ↑ apoptosis ↓	c-Maf [109], JARID2 [110]
miR-221/222	up [111]	apoptosis ↓ proliferation ↑	TIMP3 [111], PTEN [111], p27 [112]

table 1.
deregulated miRNAs in lung cancer with their experimentally validated targets (incomplete list)

2.2.6. miRNA involvement in cell cycle regulation

Increasing evidence shows that miRNAs regulate many components of the cell cycle machinery. The cell cycle can be divided into four distinct phases (figure 4): S phase (synthesis phase), M phase (mitosis phase) and two gap phases G_1 and G_2.

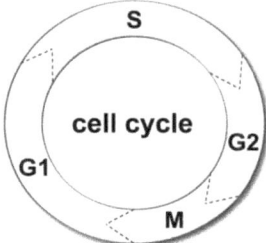

figure 4. cell cycle phases

Fully differentiated cells may encounter a prolonged quiescent state, called G_0 phase, in which they remain non-proliferative for a long period of time, awaiting environmental signals to re-enter G_1 phase. Three main restriction points, namely G_1/S, G_2/M and metaphase checkpoints are responsible for cell cycle control. Progression through these cell cycle checkpoints depends on the tightly regulated expression of different genes including RAS, PI3K/PTEN, as well as components of the Retinoblastoma pathway, cyclin-CDK complexes or cell cycle inhibitors of the INK4 or Cip/Kip families.

miRNAs have emerged as important regulators of different components of the cell cycle machinery, adding an additional layer of complexity to the coordinated regulation of the cell cycle. One of the first and maybe the most prominent example is *let-7*, an important regulator of RAS. Microarray analysis and reporter assays identified numerous additional genes involved in G_1/S or G_2/M transition including *CDK6, CDC25A* and *Cycin D2* (*CCND2*) as direct targets of *let-7* [115]. Several other components of important signalling pathways involved in cell cycle control, mainly in G_1 to S transition, have joined the list of confirmed targets of miRNAs (figure 5) [123].

Many miRNAs with oncogenic properties exert their function through the inhibition of cell cycle inhibitors (e.g. p27, p21, p16) while tumour suppressing miRNAs act on positive regulators of the cell cycle (Myc, CCND1, CDK4/6).

The fact that many miRNAs including *miR-34a, let-7* or the *miR-17-93* cluster regulate multiple cell proliferation targets (figure 5), indicates that single miRNAs may regulate cellular pathways at multiple steps.

Introduction

The complexity of such regulatory mechanisms is extended by the fact that some miRNAs form part of integrated networks with regulators of the cell cycle resulting in regulatory loops. For example, E2F induces the expression of *miR-17-92* which, in turn, down-regulates E2F [49]. An inverse situation has been described for *let-7* and *miR-34*: both miRNAs are able to down-regulate Myc, while the latter protein blocks transcription of either one of these miRNAs [80]. Both regulatory loops are linked, since Myc is able to induce expression of *miR-17-92* and thereby modulate expression of E2F (figure 5). Interestingly, Myc also directly enhances the expression of CCND1, which in turn controls *miR-17-92* expression in a regulatory feedback loop [124].

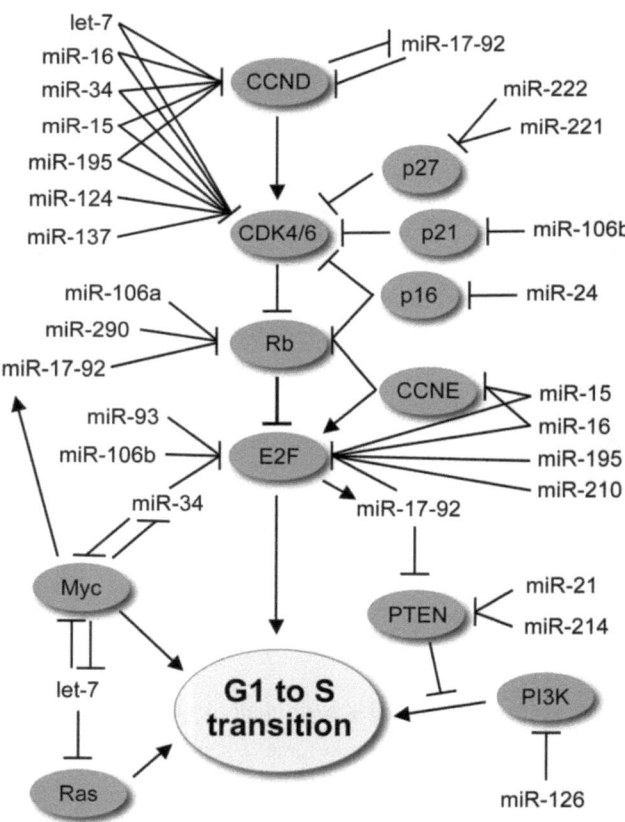

figure 5. integrated network of miRNAs that regulate G_1 to S transition of the cell cycle

Furthermore, recent findings suggest that many individual components of the cell cycle machinery are able to directly control miRNAs expression [48, 49, 124, 125]. In addition, both the stability and subcellular localisation of miRNAs may be regulated in a cell cycle dependent manner. For example *miR-29a* and *miR-29b*, encoded by a single primary transcript, are differentially expressed throughout the cell cycle. *miR-29b* is expressed at low levels except during mitosis, whereas *miR-29a* is expressed constitutively. Interestingly low level expression of *miR-29b* in all cell cycle phases except mitosis is due to rapid and selective degradation [126].

In proliferating cells, the 3'UTR of many genes is shortened or switched to an alternative 3'UTR. As a consequence this mechanism leads to an altered regulation of several genes by miRNAs during proliferation [127]. In addition, depending on proliferative status of a cell, miRNAs can also be regulated by the differential expression or activation of regulatory proteins. For example proteins including HuR and Dead End can bind to sites next to miRNA binding sites, preventing the access of miRNAs to their target mRNA (reviewed in ref. [128]). Furthermore, GW182, FXR1 and SMAD interact with RISC or the microprocessor complex and thereby can stimulate or reduce miRNA processing [42, 129, 130]. Finally, miRNAs can even switch from repression to activation of their targets, during the cell cycle [42]. Altogether the interaction networks between miRNAs and cell cycle regulatory pathways are inherently complex.

2.2.7. miR-15a/16

| hsa-miR-15a | 5' UAGCAGCACAUAAUGGUUUGUG 3' |
| hsa-miR-16 | 5' UAGCAGCACGUAAAUAUUGGCG 3' |

figure 6. sequences of mature *miR-15a* and *miR-16*, seed sequences are highlighted.

The *miR-15a / miR-16-1* cluster is located at chromosome 13q14, within the intron 4 of a gene with unknown function, called *dLeu2*. *miR-15a* and *miR-16* share the same seed sequence and therefore likely regulate the same set of target genes. They are ubiquitously expressed in normal tissue, with the highest levels in normal CD5+ lymphocytes.

In 2002, the *miR-15a/16-1* cluster was identified as potential cancer gene responsible for the pathogenesis of chronic lymphatic leukaemia (CLL) [131]. This finding provided the first indication that miRNA genes might be important for tumourigenesis. Since then, the role of miR-15a and miR-16 for the pathogenesis of various cancer types has been extensively studied. Several research groups including our group have identified direct targets of miR-15a/16 and analysed their role in tumourigenesis. To date, the miR-15a/16 cluster is one of the most prominent example of miRNAs acting as tumour suppressors. Down-regulation of these miRNAs has been reported in various tumour types including CLL [131], mantle cell lymphomas [132], multiple myelomas [133], prostate cancers [134], lung cancer [83] and in pituitary adenomas [135], suggesting that this miRNA cluster could be a general "hot spot" involved in cancer transformation. Ectopic expression of miR-15a/16 led to the up-regulation of 265 genes and to the down-regulation of 3307 genes in CLL cells [136], clearly supporting the finding that one single miRNA can have far-reaching effects on the overall function of a cell.

Cimmino et al. demonstrated that miR-15a/16 expression is inversely correlated to Bcl-2 expression and that Bcl-2 repression by these miRNAs induces spontaneous apoptosis in a leukemic cell line model [86]. In a colon carcinoma cell line, however, the same miRNAs coordinately regulate cell cycle progression [137], suggesting that these miRNAs may have cell type-specific functions. Their involvement in cell cycle control was further supported by our work [83] and the recent work by Chen et al. [138], Liu et al. and [85], Bonci et al. [84]. New direct targets including Cyclin D1 (CCND1), Cyclin D2 (CCND2), Cyclin D3 (CCND3), CDK6, Cyclin E1 (CCNE1), E2F3 were identified. Consistent with these findings, the transition from G_1 to S phase is regulated by these miRNAs. Whether miR-15a/16 induces apoptosis or/and cell cycle arrest presumably depends on the expression levels of the respective miR-15a/16 target proteins in a specific cell type.

The tumour suppressor function of miR-15a/16 has been addressed in vitro and in vivo, where expression of those miRNAs inhibits cell proliferation, promotes apoptosis and suppresses tumourigenicity in different cell types. In vivo knock-down of miR-15a and miR-16-1 has been shown to result in prostate hyperplasia associated with CCND1 and WNT3A up-regulation, indicating that the loss of these miRNAs could be a relevant event in prostate carcinogenesis [84].

Various mechanisms including loss of heterozygosity, germline and somatic mutations, epigenetic regulations and defects in the miRNA biogenesis machinery contribute to the deregulation of miR-15a/16 expression in different cancer types. However, the precise mechanisms that lead to the down-regulation of miR-15a/16 are complex and remain mostly unknown.

2.2.8. miR-34

figure 7. sequences of mature *miR-34a*, *miR-34b** and *miR-34c*, seed sequences are highlighted.

The *miR-34* family comprises three processed miRNAs that are encoded by two different genes: *miR-34a* is located at chromosome 1p36, whereas *miR-34b** and *miR-34c* map to chromosome 11q23 and share a common primary transcript. All three miRNAs, like *miR-15a/16*, are ubiquitously expressed in normal tissue with varying levels in different tissue types [94, 139]. *miR-34* acts as tumour suppressor: reduced expression has been observed in different types of cancer [96, 140-143] including lung cancer. Moreover, low levels of *miR-34a* correlate with a high probability of relapse of NSCLC [91].

Recently, the *miR-34* family was shown to form part of the p53 network [93, 97]: the expression of *miR-34a* is directly induced by p53 in response to DNA damage or oncogenic stress. Depending on cell type, the ectopic expression of *miR-34a* induces cell cycle arrest in the G_1 phase, senescence or/and apoptosis [93, 94]. As cell cycle arrest and apoptosis are common end points of the p53 pathway, *miR-34* family members may be potent mediators of the p53 signal. Furthermore, *miR-34* may contribute to the fine-tuning of the p53 response by targeting p53-induced mRNAs and thereby may prevent an uncontrolled, irreversible response to p53 activation. Ectopic over-expression of different members of the *miR-34* family in various cell lines resulted in the down-regulation of hundreds of putative *miR-34* targets. Several research groups showed that *miR-34a*, like *miR-15a/16*, directly controls the expression of CCND1, CDK4, CDK6, E2F3 and Bcl-2 but, in contrast to *miR-15a/16*, regulates the expression of SIRT-1, c-Myc, n-Myc, and c-Met [92, 93, 95, 96, 144, 145].

Down-regulation of *miR-34* is caused by inactivating mutations of *p53*, genetic alterations such as point mutations, chromosomal translocations, and/or deletions of regions encoding the *miR-34* genes [56]. In addition, reduced expression of *miR-34a* may be due to aberrant CpG methylation [139] of the *miR-34a* promotor.

2.2.9. miRNAs as diagnostic markers and therapeutics

Our understanding of microRNA expression patterns which may potentially be used as biomarkers for diagnosis, prognosis, personalized therapy, and disease management, is just starting to emerge. Recent reports suggest that expression profiling of a few hundred miRNAs has a much better predictive power for tumour classification than profiling several tens of thousands of mRNAs [66]. Thus, miRNA expression profiles could be used for the classification of poorly differentiated tumours of unknown origin. Moreover, compared to mRNAs, miRNAs are very stable and can be reliably quantified from sputum, serum, formalin-fixed paraffin-embedded tissue and very small needle biopsies.

Several microRNAs have been reported to be associated with clinical outcome and response to chemotherapy, suggesting that comparative miRNA expression studies might reveal new predictive markers for therapy, disease relapse and metastasis. Furthermore, single nucleotide polymorphisms (SNPs) in miRNA sequences [146, 147] as well as in miRNA-binding sites [74, 148] can increase cancer-risk and therefore may serve as novel biomarkers.

The fact that miRNAs are frequently deregulated in many diseases including cancer, heart failure, neurological disorders and viral infections opens a new field of potentially powerful novel therapeutic approaches. Therapeutic strategies for the treatment of human diseases based on modulation of miRNA activity have gained much attention in the past few years [149-154]. For example several authors demonstrated that miRNAs can successfully modulate the sensitivity of cancer cells to well known chemotherapeutic agents (reviewed by Ref. [155]). Thus, using miRNAs in combination with existing therapeutic strategies may potentiate the effect of current cancer treatments.

Two strategies of applications of miRNAs can be envisaged:

One strategy is to inhibit over-expressed miRNAs by using miRNA antagonists, such as anti-miRs, locked-nucleic acids (LNA), or antagomirs. These miRNA antagonists are oligonucleotides with sequences complementary to the endogenous miRNA. They carry chemical modifications that enhance their stability and the affinity to the target miRNA. Small molecule inhibitors specific for certain miRNAs are also being developed to inhibit miRNA function.

The second strategy (miRNA replacement), involves the reintroduction of chemically modified tumour-suppressor miRNA mimics to restore a the endogenous level of specific miRNAs.

Targeting multiple genes involved in several pathways relevant for human diseases using a miRNA-based therapy seems to be a very powerful approach. However, it also raises concerns about potential toxicity as a consequence of therapeutic delivery of miRNA mimics, since this will lead to increased levels of miRNAs in normal cells. Toxic effects might also be a result of overloading RISC with exogenous miRNAs as this may lead to a loss of function of endogenous miRNAs, required for normal cellular functions.

However, despite these concerns, in vivo evidence for toxicity induced by miRNAs is still lacking. On the contrary, there are a few very promising candidates of miRNA-based therapies as indicated by experiments performed in mice and non-human primates. In those studies, the delivery of therapeutic miRNAs failed to reveal adverse side effects and suggest that the delivery of miRNAs to normal tissues was well tolerated [118, 119, 154, 156]. Interestingly, a recent publication describes a new formulation that allows the reintroduction of miRNAs, depleted in cancer cells, in order to reactivate cellular pathways that drive a therapeutic response [118]. The authors demonstrated that if administrated intravenously, formulated *miR-34a* blocked tumour growth in a mouse model of NSCLC.

Even more exciting is the announcement by the biopharmaceutical company Santaris Pharma this year, that the first miRNA targeted drug therapy, miravirsen (SPC3649), reaches phase II clinical trials. Miravirsen is a specific inhibitor of *miR-122*, a liver specific miRNA that is required for the replication of Hepatitis C virus. In addition to miravirsen, Santaris Pharma is developing other drugs targeting microRNAs, some in collaboration with other companies like miRagen Therapeutics (cardiovascular diseases), Shire plc (rare genetic disorders), Pfizer (undisclosed therapeutic areas), GlaxoSmithKline (viral disease) and Enzon Pharmaceuticals (oncology).

3. Aim of the project

MicroRNAs are negative regulators of gene expression at the post transcriptional level that are frequently involved in tumourigenesis. The aim of the present work was to investigate the role of different microRNAs in the tumourigenesis of NSCLC.

At the time I started with my thesis, it was known that *miR-15a* and *miR-16* are implicated either in cell cycle control or apoptosis depending on cell system, but little information was available about their direct targets and their physiological role in solid tumours.

To investigate if *miR-15a/16* are important for the tumourigenesis of lung cancer, we wanted to assess if *miR-15a/16* are dysregulated in those tumours. Because quantification of RNAs is challenged by extensive fragmentation and modification of nucleic acids during formalin fixation, we first established conditions to quantify miRNAs from formalin-fixed paraffin-embedded (FFPE) tissues. Consequently, we collected FFPE lung cancer tissues and corresponding normal tissue by laser capture microdissection and compared them for miRNA expression and the deletion of the *miR-15a/16* locus. In parallel, we wanted to investigate, if the dysregulation of *miR-15a/16* may affect the expression of potential target genes.

In a next step we analysed the molecular mechanism underlying *miR-15a/16* regulation in NSCLC. To investigate the cellular phenotypes triggered by *miR-15a/16*, they were over-expressed in NSCLC cell lines. In parallel, luciferase constructs were made containing predicted target sequences downstream of a luciferase reporter gene to identify direct targets of *miR-15a/16*. Using these luciferase constructs we analysed if physiological concentrations of miRNAs in NSCLC cells are able to down-regulate targets of *miR-15a/16*. This type of experiments is important because it is unclear for most miRNA-targets if they are also regulated at physiological concentrations of endogenous miRNAs in specific cell types. In addition, we analysed molecular mechanisms by which NSCLC tumour cells can escape the growth inhibitory signals of *miR-15a/16*.

When I started with the second part of this thesis, much more information was available about miRNAs involved in cell cycle control and apoptosis. Although it was becoming clear that many miRNAs share overlapping functions and form part of integrated networks, little information was available how miRNAs interact with each other to control biological processes. Thus, using the example of *miR-15a/16* and the closely related *miR-34a*, we wanted to investigate how miRNAs with overlapping functions affect biological processes in a combinatorial mode. It has been shown previously that *miR-34a* and *miR-15a/16* are involved in similar processes, since they are able to regulate a similar set of target genes. We analysed if *miR-34a* and *miR-15a/16* are involved in the same regulatory network for cell cycle control in NSCLC cells. In addition, to investigate, how *miR-15a/16* and *miR-34a* interact with each other, NSCLC cells were co-transfected with both miRNAs and analysed for cell cycle progression and target gene repression. In parallel, the same tumour samples that we used for the analysis of *miR-15a/16* expression were also analysed for *miR-34a* expression, which allowed us to asses if *miR-15a/16* and *miR-34a* are co-regulated in NSCLC.

The understanding of the complex mechanisms, how tumour cells can escape the tumour-suppressor function of miRNAs like *miR-15a/16* and *miR-34a*, will enhance the knowledge of the involvement of miRNAs in tumourigenesis. The fact that miRNAs are frequently dysregulated in cancer opens a new field for the treatment of human cancer. In particular, miRNAs involved in cell cycle control and apoptosis may be interesting candidates for miRNA based cancer therapy.

4. Results

Manuscript 1

miR-15a and *miR-16* are implicated in cell cycle regulation in a Rb-dependent manner and are frequently deleted or down-regulated in non-small cell lung cancer

Nora Bandi[1], Samuel Zbinden[1], Mathias Gugger[1], Marlene Arnold[1], Verena Kocher[1], Lara Hasan[1,2], Andreas Kappeler[1], Thomas Brunner[1] and Erik Vassella[1,3]

Cancer Research 2009;69:5553-9.

[1]Institute of Pathology, University of Bern, Bern, Switzerland
[2]Present address: BÜHLMANN Laboratories AG, CH-4124 Schönenbuch, Switzerland

[3]Corresponding author: E. Vassella
                Institute of Pathology
                University of Bern
                Murtenstrasse 31
                CH-3010 Bern
                Switzerland

                phone: 0041-31-632 9943
                fax: 0041-31-381 8764
                Email: erik.vassella@pathology.unibe.ch

Running title: microRNA and tumourigenesis of lung cancer
Key words: microRNA, non-small cell lung cancer, cell cycle control, cyclin, retinoblastoma

Manuscript 1

miR-15a and *miR-16* are implicated in cell cycle regulation in a Rb-dependent manner and are frequently deleted or down-regulated in non-small cell lung cancer

Nora Bandi, Samuel Zbinden, Mathias Gugger, Marlene Arnold, Verena Kocher, Lara Hasan, Andreas Kappeler, Thomas Brunner and Erik Vassella

Institute of Pathology, University of Bern, Bern, Switzerland

Abstract

MicroRNAs (miRNAs) are negative regulators of gene expression at the post-transcriptional level that are involved in tumourigenesis. Two miRNAs, *miR-15a* and *miR–16*, that are located at chromosome 13q14, have been implicated in cell cycle control and apoptosis, but little information is available about their role in solid tumours. To address this question, we established a protocol to quantify miRNAs from laser capture microdissected tissues. Here we demonstrate that *miR-15a*/*miR-16* are frequently deleted or down-regulated in squamous cell carcinomas and adenocarcinomas of the lung. In these tumours expression of *miR-15a*/*miR-16* inversely correlates with the expression of cyclin D1. In non-small cell lung cancer (NSCLC) cell lines, cyclin D1, cyclin D2 and cyclin E1 are directly regulated by physiological concentrations of *miR–15a*/*miR-16*. Consistent with these results, overexpression of these miRNAs induces cell cycle arrest in G1/G0. Interestingly, H2009 cells lacking Rb are resistant to *miR–15a*/16-induced cell cycle arrest, while re-introduction of functional Rb resensitizes these cells to miRNA activity. In contrast, down-regulation of Rb in A549 cells by RNA interference confers resistance to these miRNAs. Thus, cell cycle arrest induced by these miRNAs depends on the expression of Rb, confirming that G1 cyclins are major targets of *miR-15a*/*miR-16* in NSCLC. Our results indicate that *miR-15a*/*miR-16* are implicated in cell cycle control and likely contribute to the tumourigenesis of NSCLC.

Introduction

Lung cancer is the leading cause of cancer-associated death and is responsible for more deaths than the next three most common tumours combined (breast, prostate and colon) (1). The high mortality rate of this disease entity is primarily due to the fact that it is only diagnosed at an advanced stage. Lung cancer comprises several histological types including small cell lung cancer (SCLC) and non-small cell lung cancer (NSCLC); the latter can be further subdivided into two major types, squamous cell carcinoma and adenocarcinoma (2). Squamous cell carcinomas usually arise from the major bronchi while adenocarcinomas arise from distant airway bronchioles and alveoli.

Lung cancer is characterized by multiple genetic changes affecting different oncogenes or tumour suppressor genes involved in cell cycle control, DNA repair and apoptosis (2). The most common alterations are overexpression of cyclin D1 (CCND1) and Bcl2, mutations of KRAS or members of the ERBB family, and mutations or inactivation of Rb, p16^{INK4a} and TP53 (2). Altered expression of marker genes in neoplastic tissues may be due to genetic alterations or promoter methylation, but in many cases the mechanism of dysregulation is unknown.

MicroRNAs (miRNAs) constitute a novel class of regulatory molecules at the post transcriptional level that are involved in tumourigenesis (3). These short RNAs of 19-25 nucleotides play key roles in a wide variety of biological processes including cell fate specification, proliferation, cell death and energy metabolism (4). Several miRNAs have been identified, which are directly involved in tumourigenesis of lung cancer and which are considered as prognostic markers for overall survival or predictive markers for chemotherapy (5, 6).

miR-15a and *miR-16*-1 are located at chromosome 13q14, a region which is deleted in 68% of chronic lymphocytic leukaemia (CLL) (7). Cimmino et al. demonstrated that *miR-15a/miR-16* expression is inversely correlated to Bcl2 expression and that Bcl2 repression by these miRNAs induces spontaneous apoptosis in a leukemic cell line model (8). In a colon carcinoma cell line, however, the same miRNAs coordinately regulate different mRNA targets including CDK6, CARD10, CDC27 and C10orf46 that act in concert to control cell cycle progression (9), suggesting that these miRNAs may have cell type-specific functions. The spectrum of potential targets of *miR-16* was further expanded by the recent work of Chen et al. (10), Liu et al. (11) and Bonci et al. (12) who demonstrated that G1 cyclins are also regulated by this miRNA.

To investigate if *miR-15a/miR-16* are important for the tumourigenesis of lung cancer we collected tissues from squamous cell carcinomas and adenocarcinomas and corresponding normal tissues by laser capture microdissection and compared these tissues for miRNA expression. We report that *miR-15a/miR-16* are deleted or down-regulated in the majority of NSCLC. In addition, we investigated the molecular mechanism underlying *miR-15a/miR-16* regulation in different NSCLC cell lines. We demonstrate that

Manuscript 1

miR-15a/miR-16 are able to induce cell cycle arrest in G1/G0 in an Rb-dependent manner and that G1 cyclins are physiological targets of these miRNAs. We propose two alternative pathways by which NSCLC cells escape *miR-15a/miR-16*-induced cell cycle arrest: (i) downregulation of *miR-15a/miR-16* and (ii) inactivation of the Rb gene.

Materials and Methods
Cell lines and culture conditions

The NSCLC cell lines Calu-1, Calu-6, A549, H2009, H1299 and H358 were obtained from the American Type Culture Collection, Rockville, MD. All cell lines were cultured in Iscove's modified Dulbecco's medium supplemented with 2 mM L-alanyl-L-glutamine, 1% penicillin/streptomycin and 5% foetal bovine serum (Sigma, St. Louis, MO, USA) at 37°C and 5% CO_2.

Constructs

Luciferase reporter constructs were generated containing a firefly luciferase gene cloned between the *Hind*III and *Xba*I sites of pcDNA3.0 (Invitrogen, Paisley, UK). DNA fragments encompassing the 3' UTR of cyclin D1 (CCND1), cyclin D2 (CCND2) and cyclin E1 (CCNE1), respectively, were amplified from genomic DNA and cloned into the *Xba*I site of the luciferase construct. Luciferase plasmids containing one or three copies of a *miR-15a/miR-16* target site from *CCND1*, *CCND2* or *CCNE1*, respectively, were obtained by cloning double-stranded oligonucleotides into the *Xba*I site of the luciferase plasmid (suppl. Table 1 and suppl. Fig. 1). pcDNA3 3xmiR15a/16 was constructed by cloning three copies of the *miR-15a/miR-16* locus in tandem into pcDNA3.0. Rc/CMV cyclinD1 and Rc/CMV cyclinE expression constructs (13) were obtained from Addgene Inc (Cambridge, MA). The *miR-15a/miR-16* target sites were deleted from Rc/CMV cyclin E1 by PCR amplification and subcloning into pcDNA3.0 giving rise to Rc/CMV cyclinEΔ3'UTR. Primers used for amplification are indicated in suppl. Table 1.

Transfection

Cells were seeded in culture flasks 24 hrs prior to transfection. Co-transfections with plasmid DNA were performed using Effectene reagent (Qiagen, Hilden, Germany), all other transfections were performed using HiPerFect (Qiagen). Transfection was performed using 10-20 nM of a mixture of equal amounts of hsa pre-*miR-15a* and hsa pre-*miR-16* precursors or pre-miR miRNA precursor control 1 (Ambion, Cambridge, UK), 60 nM Rb-kd siRNA (siGENOME SMARTpool, Dharmacon, Lafayette, CO), 60 nM siCONTROL non-targeting Pool 2 (Dharmacon), 100 nM of a mixture of equal amounts of anti-*miR-15a* and anti-*miR-16* inhibitors or anti-miRNA inhibitor control (Ambion, Cambridge, UK), pCMV-Rb (14) or empty control plasmid, firefly luciferase constructs and Renilla luciferase reporter plasmid pGl4.74 (Promega, Madison, WI, USA) or expression constructs as described above. Transfection efficiency of short RNAs and plasmid DNA was monitored using siGloGreen transfection indicator (Dharmacon) or an RFP-expression plasmid, respectively.

Luciferase activity assay, cell cycle analysis and apoptosis assay

Luciferase activity assays were performed 48 hrs after transfection using a dual-luciferase reporter assay system (Promega) and an Infinite 200 reader (Tecan, Männedorf, Switzerland).

Cell cycle analysis was performed essentially as described (9). Cells were analysed using an LSR II flow cytometer (BD Biosciences, San Jose, CA, USA) and FlowJo 8.8.4 software (Tree Star, Ashland, OR).

For apoptosis assays, floating and adherent cells were harvested 24-72 hrs after transfection, combined and washed with PBS. Annexin V (1:100) either alone or in combination with 10 µg/ml propidium iodide (Sigma) was added to the cells and samples were analysed within 30 min after staining. Quantification of fluorescence was performed by flow cytometry as described above.

RNA isolation and real-time PCR

Total RNA was extracted from cultured cells using the miRVana RNA isolation kit according to the manufacturer's instructions (Ambion). TaqMan miRNA assays (Applied Biosystems, Foster City, CA) were used to quantify the expression of mature hsa-*miR-15a/miR-16*. Reverse transcription was performed using the TaqMan MicroRNA reverse transcription kit. Reverse transcription of mRNAs was performed using TaqMan reverse transcription reagents (Applied Biosystems). Quantitative PCR (qPCR) of CCND2 was performed using a TaqMan assay (Applied Biosystems); all other amplifications were performed using Quantitect Primer assays (Qiagen). Quantitative PCR was performed in a Real-Time PCR system 7500 (Applied Biosystems). Gene expression of miRNAs was calculated relative to RNU48 or RNU6B; mRNA levels were normalized to the level obtained for GAPDH, hb2m, HPRT-1 or RPL13A, respectively. Changes in expression were calculated using the ΔΔCt method.

Western blot analysis and immunohistochemistry

Proteins were separated by SDS-PAGE and Western blotting was performed using PVDF membrane (Millipore, Billerica, MA, USA). Unspecific sites were blocked in 5% non-fat dry milk or 5% BSA at RT for 1 hr. Antibodies directed against CCND1 (clone DCS-6, Dako, Glostrup, Denmark; diluted 1:100), CCNE1 (clone 13A3, NeoMarkers, Fremont, CA, USA; 1:200), Rb (clone 3C8, QED Bioscience, San Diego, CA, USA; 1:1000), phospho-Rb (Ser807/811, Cell Signaling Technology, 1:1000) and α-tubulin (clone B512, Sigma; 1:5000) were used. Secondary antibody was goat anti-mouse-HRP or goat anti-rabbit-HRP (Biorad, Reinach, Switzerland) used at 1:5000 or 1:7000, respectively.

For immunohistochemistry, 4 µm formalin-fixed, paraffin-embedded sections were treated essentially as described (15). Anti-Bcl2 (Clone 124, Dako), anti-CCND1, anti-CCNE1 and anti-Rb antibodies were used at 1:30, 1:25, 1:25 and 1:100 dilutions, respectively. Mouse IgG1 (Dako, 1:20) was used as a negative control. Sections were incubated with EnVision+ system (labelled polymer HRP anti mouse, Dako) for 30 min at room temperature, developed in 3,3'-Diaminobenzidin (Sigma) for 8 min and counterstained with haematoxylin.

Laser capture microdissection and LOH analysis

FFPE sections were deparaffinised and stained with methylgreen (Merck, Darmstadt, Germany). Approximately 1000 tumour cells were captured onto an adhesive cap using a PALM Microbeam (PALM, Zeiss, Bernried, Germany). The dissected material was resuspended in digestion buffer (RecoverAll, Ambion). 120 ng tRNA were added to the samples and incubated for 10 min at 90°C. All subsequent purification steps were performed following the manufacturer's instructions except that the DNase treatment step was omitted. Loss of heterozygosity (LOH) analysis was performed as described (7) using a genetic analyser (ABI Prism 3100 Avant, Applied Biosystems) and the GeneMapper software 4.0.

All experiments using human specimens were done according to the ethical guidelines of the Institute of Pathology, University of Bern, and were reviewed by the institutional review board.

Statistics

Statistical differences were calculated using unpaired two-tailed student's t-test. A probability of p ≤0.05 was considered statistically significant. Statistical significance of inverse correlation was calculated by the N-1 chi-square test (two by two table Pearson's analysis (16).

Results

miR-15a and *miR-16* are frequently down-regulated in lung cancer

It has previously been shown that *miR-15a/miR-16* can induce apoptosis or cell cycle arrest depending on the cell line (8-11). To address their role in NSCLC, we analysed 11 adenocarcinomas and 12 squamous cell carcinomas of the lung from the archive of the Institute of Pathology at the University of Bern. Quantification of RNAs is challenged by extensive fragmentation and modification of nucleic acids during formalin fixation (17). To address this problem, nucleic acids were subjected to a heat treatment prior to RNA extraction, which reverses methylol groups introduced during formalin fixation. Under these conditions, fresh and FFPE tissues gave rise to similar *miR-16* levels as indicated by qPCR (suppl. Fig. 2A). The quality of miRNAs may also depend on the post-operative time period prior to fixation. Experimental tissue samples were left at room temperature for up to five hrs prior to fixation which, however, did not affect *miR-16* quantification (suppl. Fig. 2A). In addition, detection of *miR-16* was linear over a wide concentration range (suppl. Fig. 2B) indicating that *miR-16* can be quantified accurately from formalin-fixed laser-captured microdissected material.

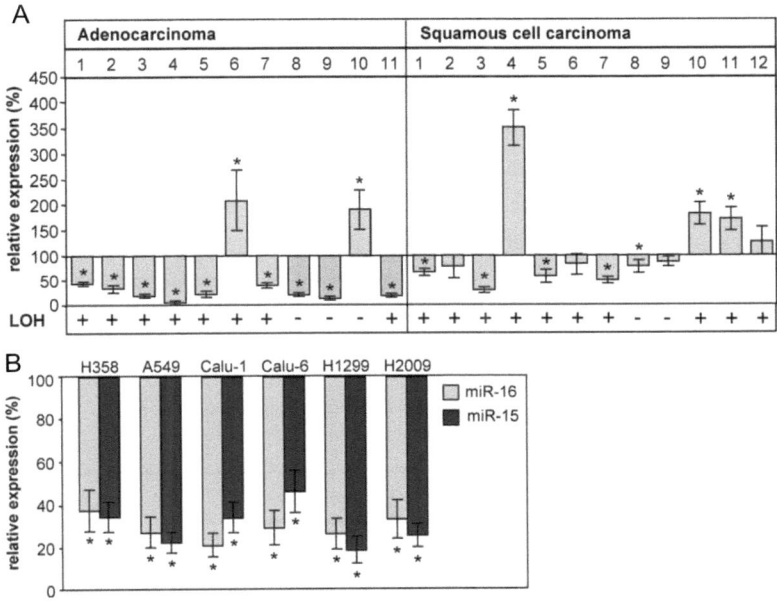

Figure 1. *MiR-15a* and *miR-16* are frequently deleted or down-regulated in non-small cell lung cancer. (A) *MiR-16* levels in adenocarcinomas and squamous cell carcinomas relative to matched normal tissues (n=3). The LOH status for the presented samples is shown as: -, heterozygosity; +, LOH. Tumours with no informative microsatellite locus were excluded. (B) *MiR-15a* and *miR-16* levels in NSCLC cell lines relative to normal lung tissue (n=3). *, p<0.05.

Tumour tissues from adenocarcinomas and squamous cell carcinomas were compared to matched normal tissue from alveolar or bronchial epithelium, respectively, which had been microdissected from the same slide. *MiR-16* was down-regulated in 82% (9 of 11) of adenocarcinomas (p<0.001) and in 67% (8 of 12) of squamous cell carcinomas (in 5 of which down-regulation was statistically significant, p<0.05) (Fig 1A). Interestingly, *miR-16* was significantly upregulated in 26% of tumours. *MiR-15a* gave rise to a similar expression pattern as *miR-16* (data not shown), but was more difficult to detect due to its low abundance. Both miRNAs were also significantly down-regulated in the NSCLC cell lines Calu-1, Calu-6, A549, H2009, H1299 and H358 (Fig. 1B).

In the majority of cases, down-regulation of *miR-15a/miR-16* was associated with a deletion of one allele of the *miR-15a/miR-16* locus. This was based on the finding that two microsatellites, D13S273 and D13S272, flanking the *miR-15a/miR-16* locus revealed a loss of heterozygosity (LOH) in 73% (8 of 11) of adenocarcinomas and in 83% (10 of 12) squamous cell carcinomas (Fig. 1A).

MiR-16 expression inversely correlates to cyclin D1 in NSCLC

To investigate if dysregulation of *miR-16* may affect the expression of potential targets, CCND1, CCNE1 and Bcl2 were analysed by immunohistochemistry. Analysis of the same tumour area on serial tissue sections revealed an inverse correlation in the expression of *miR-16* and CCND1 in 19 out of 23 samples (Table 1 and suppl. Fig. 3) suggesting that CCND1 is a physiological target. In contrast, no inverse correlation was observed for CCNE1 or Bcl2. Based on these results, we cannot exclude that the latter genes are targets, since additional layers of gene regulation may exist which contribute to CCNE1 and Bcl2 expression.

Table 1: Expression of CCND1, CCNE1 and Bcl2 relative to *miR-16* in adenocarcinomas and squamous cell carcinomas of the lung

	Protein expression	*miR-16* expression*		n	p
		low	high		
CCND1	-	1	3	23	0.02
	+	16	3		
CCNE1	-	10	7	23	0.3
	+	5	1		
Bcl2	-	12	3	23	0.4
	+	5	3		

*Number of tumours with reduced or increased levels of *miR-16* relative to normal tissue.

Cellular phenotypes triggered by *miR-15a* and *miR-16* in NSCLC cell lines

To investigate the cellular phenotypes triggered by these miRNAs, NSCLC cell lines were co-transfected with *miR-15a/miR-16* precursors and analysed for apoptosis and cell cycle arrest. None of these cell lines underwent spontaneous apoptosis as indicated by propidium iodide staining (Fig. 2A), annexin V-staining or by morphological changes (data not shown). However, in five of six cell lines, *miR-15a/miR-16* induced cell cycle arrest in G1/G0 in a significant percentage of the cell population ($p<0.01$; Fig. 2B). This phenotype was more pronounced when cells were treated with nocodazole 24 hrs after transfection with miRNA precursors (p<0.001). In the case of A549, 33% of the cell population had undergone arrest in G1/G0 whereas only 8% of the control transfected with an unrelated RNA were in this phase of the cell cycle (Fig. 2B and 2C). Comparable results were obtained for the cell lines Calu-1, Calu-6, H1299 and H358 (Fig. 2B and suppl. Fig. 4). In marked contrast, only 2% of H2009 cells were in the G1/G0 phase upon co-transfection with *miR-15a/miR-16* precursors (Fig. 2B and 2C).

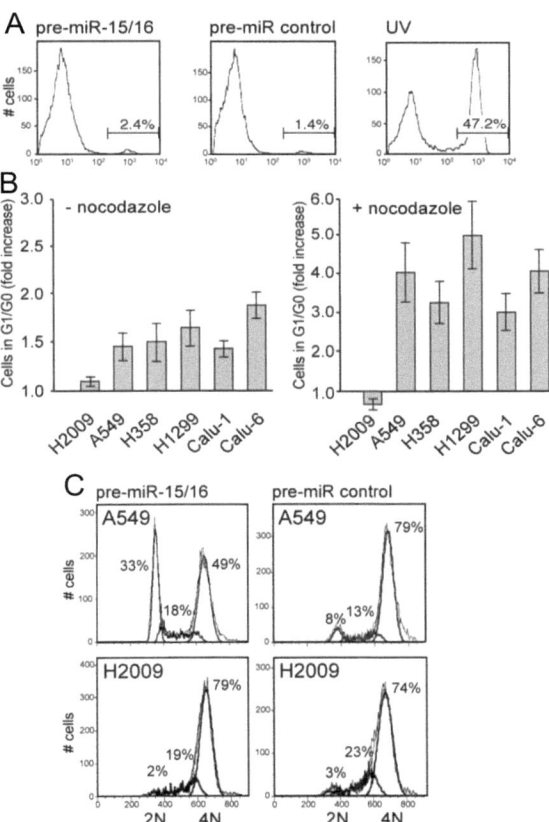

Figure 2. Phenotypic analysis of *miR-15a/miR-16* overexpression. (A) Spontaneous apoptosis. A549 cells were stained with propidium iodide 72 hrs post transfection. As a positive control, cells were treated with UV (200 mJ). (B) G1/G0 arrest. Untreated cells or cells treated with nocodazole for 20 hrs beginning at 24 hrs post transfection were analysed by flow cytometry. The increase in G1/G0 of cells co-transfected with *miR-15a/miR-16* precursors relative to cells transfected with precursor control is presented for each cell line (n=3-4). (C) DNA content distribution of A549 or H2009 cells treated with nocodazole as described above. 2N, cells having diploid DNA content; 4N, cells having tetraploid DNA content.

It can be excluded that the low percentage of cells in G1/G0 was due to a low transfection efficiency, since ~90% of H2009 cells were transfected with siGloGreen (fluorescently labelled non-functional siRNA). Thus, the cell cycle regulatory activity of *miR-15a/miR-16* is cell line-specific.

Down-regulation of endogenous G1 cyclin mRNAs by *miR-15a* and *miR-16*

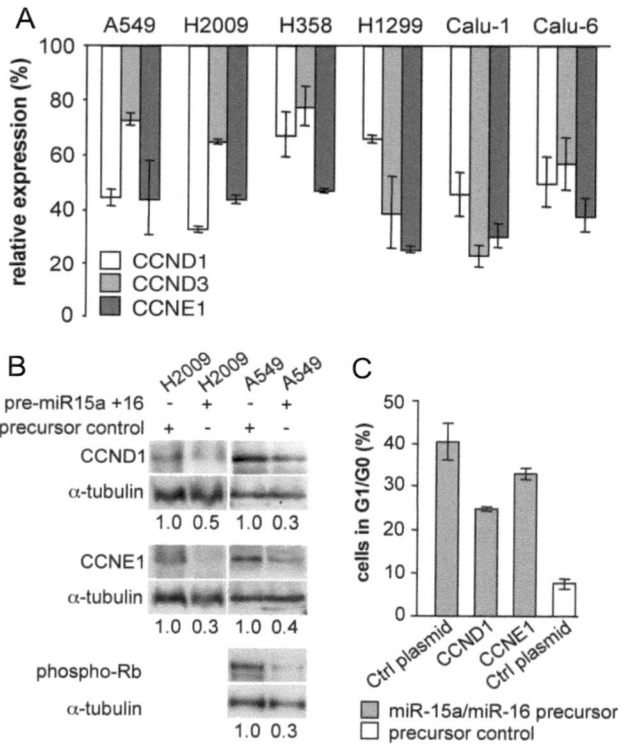

Figure 3. *MiR-15a* and *miR-16* induce down-regulation of G1 cyclin mRNAs and proteins. (A) mRNA levels. Cells were co-transfected with *miR-15a/miR-16* precursors and harvested 48 hrs post transfection. The steady-state level of G1 cyclin mRNAs was normalized to the level obtained for GAPDH. Expression values are presented as means (n=3) ±S.D. relative to the expression values obtained for the control transfected with precursor control. Comparable results were obtained when mRNAs were normalized to hb2M, HPRT-1 or RPL13A mRNAs, respectively (data not shown). (B) Protein levels. Transfected cells were analysed by Western blot using antibodies as indicated in the figure. Protein levels were normalized to a-tubulin and presented relative to the levels obtained for the control. (C) Overexpression of CCND1 and CCNE1 partially restores the effect of *miR-15a/miR-16*. A549 was co-transfected with miRNA precursors and expression constructs and treated for 20 hrs with nocodazole beginning 48 hrs post transfection. The percentage of cells in G1/G0 was determined by flow cytometry (n=3).

A search of the TargetScan database allowed the identification of CCND1, CCND2, CCND3 and CCNE1 as predicted targets of *miR-15a/miR-16* (suppl. Fig. 5). To analyse if *miR-15a/miR-16* are able to regulate the steady-state level of mRNA of these potential target genes, cells were co-transfected with *miR-15a/miR-16* precursors and analysed for the expression of G1 cyclin mRNAs by qPCR. CCND2 did not yield any PCR product in most of the cell lines analysed. The results of the other three G1 cyclins are shown in Fig. 3A. All cell lines transfected with *miR-15a/miR-16* precursors gave rise to reduced levels of CCND1, CCND3 and CCNE1 mRNAs relative to the control transfected with an unrelated RNA.

Co-transfection of A549 and H2009 cells with *miR-15a/miR-16* precursors also resulted in 50–70% less CCND1 and CCNE1 proteins relative to the control (Fig. 3B). Thus, both cell lines are equally able to down-regulate G1 cyclins in response to *miR-15a/miR-16*. In agreement with these results, the level of phospho-Rb was significantly reduced (Fig. 3B).

To confirm that cell cycle arrest was due to down-regulation of G1 cyclins, A549 was co-transfected with *miR-15a/miR-16*-refractory expression constructs and *miR-15a/miR-16* precursors. Overexpression of CCNE1 and CCND1 was confirmed by Western blotting (suppl. Fig. 6). As shown in Fig. 3C, co- transfection with *miR-15a/miR-16* precursors in combination with Rc/CMV cyclinD1 (p=0.003) or Rc/CMV cyclinEΔ3'UTR (p=0.04) in each case lead to a significant decrease in G1/G0 arrest as compared with cells co-transfected with *miR-15a/miR-16* precursors in combination with a control construct. Thus, both CCND1 and CCNE1 can partially rescue *miR-15a/miR-16*-induced cell cycle arrest.

G1 cyclins are physiological targets of *miR-15a* and *miR-16* in NSCLC cell lines

To investigate if G1 cyclins are regulated directly by *miR-15a/miR-16*, a series of luciferase constructs were made containing predicted target sequences (TS) from the different G1 cyclins cloned downstream of the luciferase reporter gene (Fig. 4A and suppl. Fig. 1) and co-transfected into H2009 cells together with *miR-15a/miR-16* precursors. CCND1 contains two potential target sites for *miR-15a/miR-16* (suppl. Fig. 5). A construct containing the first target site within the 3' UTR of CCND1 (Luc 1xTS) resulted in 35% luciferase activity relative to the parental Luc construct (Fig 4B). The activity obtained with this construct was comparable to a construct containing a part of the 3' UTR of CCND1 encompassing both target sites (Luc 3' UTR). A third construct containing three copies of a single target site in tandem (Luc 3xTS) gave rise to 25% activity. To confirm the specificity of the assay, mutations were introduced into the target sequence of *miR-15a/miR-16* giving rise to Luc mTS (Fig. 4A and suppl. Fig. 1). The activity obtained with this construct was almost restored to the activity obtained with the Luc construct. CCND1 and CCNE1 constructs gave rise to comparable luciferase activities (Fig. 4B).

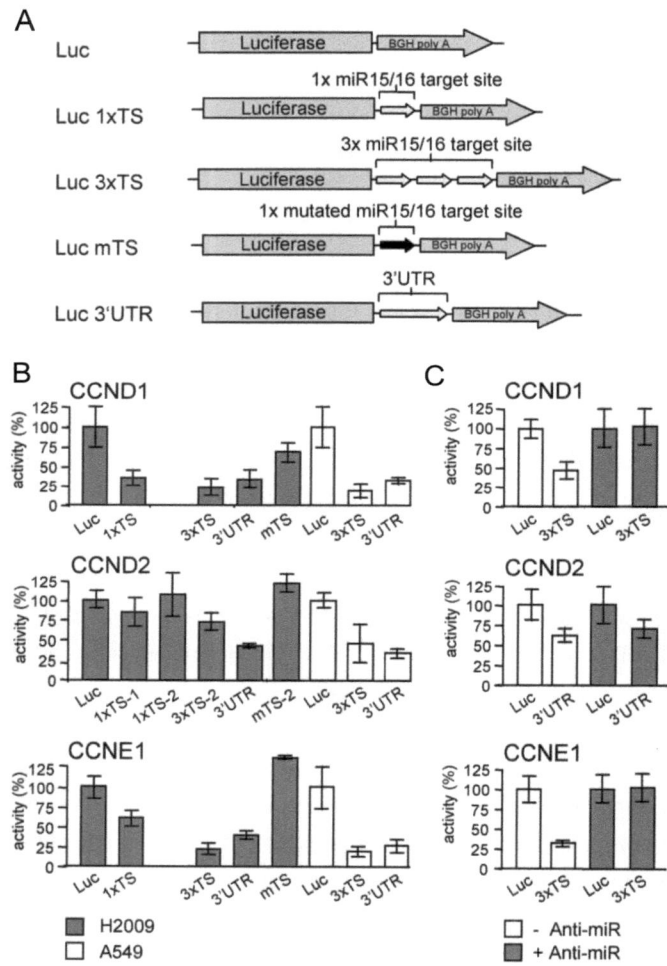

Figure 4. G1 cyclins are directly regulated by endogenous *miR-15a* and *miR-16*. (A) Luciferase constructs containing *miR-15a/miR-16* target sites (TS) from the different G1 cyclins. (B) Luciferase activity of H2009 and A549 cells co-transfected with *miR-15a/miR-16* precursors and luciferase constructs. Luciferase activity is presented relative to the activity obtained with the Luc construct (n=3). (C) Down-regulation of G1 cyclins by endogenous *miR-15a/miR-16*. Cells were transfected with luciferase constructs alone (open columns) or in combination with anti-*miR-15a* and anti-*miR-16* (grey columns) in the absence of precursor miRNAs (n=3-6).

In contrast, CCND2 seems to be less responsive to *miR-15a/miR-16*. Insertion of a single target site (TS-1 or TS-2) or three copies of one target site (TS-2) in tandem into the Luc construct did not significantly reduce luciferase activity (Fig. 4B). However, a region of the 3' UTR encompassing two target sites conferred efficient down-regulation of luciferase activity. Thus, one target site may not be sufficient for down-regulation of luciferase activity. Again, comparable results were obtained in H2009 and A549 cells (Fig. 4B).

We next investigated if G1 cyclins are regulated by endogenous *miR-15a/miR-16*. H2009 cells were transfected with the Luc 3xTS construct containing three target sites from the CCND1 or CCNE1 genes, respectively, or with the Luc 3'UTR construct containing CCND2-specific sequences. All three constructs gave rise to luciferase activities which were significantly lower than the activity obtained with the Luc construct (Fig. 4C). Relative activities were 47% (p<0.00001) for the CCND1 Luc 3xTS construct, 33% (p=0.003) for the CCNE1 Luc 3xTS construct and 62% (p=0.001) for the CCND2 Luc 3'UTR contruct. These results clearly indicate that endogenous levels of *miR-15a/miR-16* are sufficient to down-regulate G1 cyclins. In agreement with these results, anti-miRs specific for *miR-15a/miR-16* were able to completely restore the activity obtained with the Luc3xTS constructs from CCND1 and CCNE1 (p<0.002, Fig. 4C) to the level obtained with the Luc construct.

These results indicate that although *miR-15a/miR-16* are significantly reduced (Fig. 1B) they are still able to control G1 cyclins in NSCLC cell lines. To investigate if down-regulation of *miR-15a/miR-16* is a mechanism by which NSCLC cells induce overexpression of target genes, H2009 cells were co-transfected with an expression construct containing three copies of the *miR-15a/miR-16* locus and CCND1 Luc 3xTS. In these cells, the level of miRNAs was restored almost to the level obtained in normal lung tissue. These cells exhibited two times less luciferase activity relative to the control (suppl. Fig. 7), demonstrating that a reduction in *miR-15a/miR-16* activity as we observed in our lung cancer cell lines leads to overexpression of CCND1.

MiR-15a and *miR-16*-induced cell cycle arrest depends on Rb expression

Based on our results we may conclude that *miR-15a/miR-16*-induced cell cycle arrest is due to down-regulation of G1 cyclins, but this may not apply for H2009 which is resistant. H2009 cells differ from the other NSCLC cell lines by the fact that they lack Rb. Cyclin D in complexes with CDK4 and CDK6 and cyclin E in a complex with CDK2 regulate progression through the G1/S boundary of the cell cycle. These complexes phosphorylate and thereby prevent Rb from binding to E2F, which upon release, drives cells from G1 into S phase (reviewed by (18)). Thus, Rb-deficient cells no longer depend on cyclin D (19, 20) and therefore may not respond to *miR-15a/miR-16*. To test this hypothesis, Rb was expressed in H2009 cells and analysed for *miR-15a/miR-16*-induced cell cycle arrest. Co-transfection with *miR-15a/miR-16* precursors and an empty control plasmid induced cell cycle arrest in 7.4±1% of the population (Fig. 5A). The percentage of cells in the G1/G0 phase of the cell cycle increased to 23±1% upon co-transfection with an Rb expression plasmid and control precursor RNA (p=0.001), indicating that Rb *per se* is able to induce cell cycle arrest in a relatively high number of cells. However, Rb in combination with *miR-15a/miR-16* precursors induced cell cycle arrest in significantly more cells (30±3%, p=0.01) than Rb in combination with the control precursor (Fig. 5A). Thus, *miR-15a/miR-16* depend on Rb in order to exert their phenotype. The expression of Rb protein was not affected by coexpression with *miR-15a/miR-16*, but phospho-Rb was reduced by 50% (Fig. 5C).

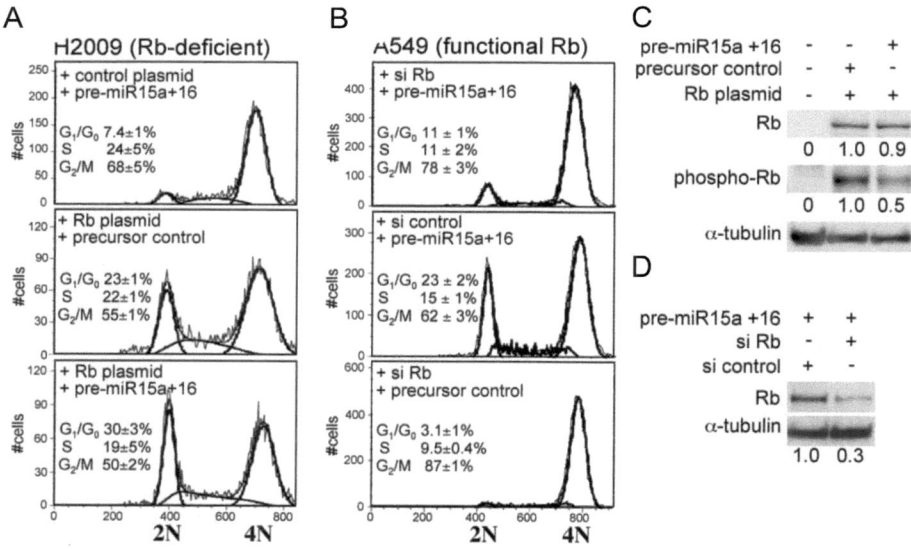

Figure 5. *MiR-15a/miR-16*-induced cell cycle arrest depends on Rb. (A,B) H2009 and A549 cells were treated with nocodazole beginning 48 hrs post transfection and analysed by flow cytometry (n=3). (C,D) Western blot analysis of H2009 (C) and A549 cells (D) subjected to the same conditions as in (A,B) using antibodies directed against Rb or phospho-Rb. Protein levels were normalized to α-tubulin.

To verify our results, the complementary experiment was performed in A549 cells in which the Rb gene was knocked down by RNA interference. These cells expressed three times less Rb protein than the control (Fig. 5D). As expected, A549 cells transfected with siRNAs against Rb were significantly more resistant to *miR-15a/miR-16*-induced cell cycle arrest (11±1% in G1/G0) than cells transfected with si control RNA (23±2% in G1/G0, p<0.001) (Fig. 5B). In contrast, Rb siRNA in combination with precursor control RNA had no effect (3±1%). In conclusion, loss of Rb confers resistance to *miR-15a/miR-16*-induced cell cycle arrest.

Discussion

Deregulated cell proliferation is a key mechanism for neoplastic progression (21). This study demonstrates that *miR-15a/miR-16* are negative regulators of cell cycle progression in NSCLC. We propose two mechanisms by which tumour cells can escape *miR-15a/miR-16*-induced cell cycle arrest, either by down-regulation of *miR-15a/miR-16* or, alternatively, by inactivation of Rb.

MiR-15a/miR-16 are expressed at reduced levels in ~70% of adenocarcinomas or squamous cell carcinomas and in six NSCLC cell lines; in the majority of cases this is associated with a LOH of the miRNA locus. However, other mechanisms may contribute to miRNA down-regulation. For example, miRNA processing is compromised owing to reduced expression of Dicer in a significant number of lung cancers (22). Transcription of the *miR-15a/miR-16* locus may also be affected. Chung et al. demonstrated that the promoter activity was reduced by PAX5 in B-cell neoplasm (23) and it is likely that a similar mechanism may exist in lung cancer. In addition, a point mutation in the *miR-16* primary transcript has been linked to diminished expression of the mature miRNA (24). Our results suggest that down-regulation of *miR-15a/miR-16* may contribute to tumour growth since this is directly coupled to increased levels of G1 cyclins.

That *miR-15a/miR-16*-induced cell cycle arrest depends on Rb was evidenced by our findings that H2009 cells lacking Rb are completely resistant to *miR-15a/miR-16*, while introduction of a functional copy of Rb into these cells renders them more sensitive. Consistent with these results, down-regulation of Rb in A549 by RNA interference confers resistance to these miRNAs. This mechanism may apply for up to 14% of squamous cell carcinomas and 33% of adenocarcinomas, which are Rb-negative (25). We found that Rb was down-regulated in 4 out of 11 adenocarcinomas (tumour #5, 6, 7 and 9) and in 3 out of 12 squamous cell carcinomas (tumour #1, 4 and 12, see Fig. 1). It is unlikely that two Rb-related proteins, p107 and p130, can compensate for a loss of Rb, since their function is distinct from that of Rb (26). In addition, these proteins are normally expressed in H2009 cells, which are refractory to *miR-15a/miR-16* (27). In conclusion, 87% (20/23) of NSCLC had either down-regulated *miR-15a/miR-16* or inactivated Rb.

Hemizygous or homozygous loss of 13q14 has been associated with numerous malignancies including CLL (7), mantle cell lymphoma (28), multiple myeloma (29), breast cancer (30) and high-grade carcinoma of the prostate (31), suggesting that these deletions are of pathogenetic significance. *MiR-15a/miR-16* are presumably the long-sought candidate tumour suppressors of this region and proved to be important regulators of Bcl2 expression in CLL (7, 8). *MiR-15a/miR-16* are also frequently down-regulated in CLL and pituitary adenomas (32). Interestingly, both miRNAs are frequently up-regulated in cervical cancer (33). Since these cells normally express an inactive form of Rb, we may conclude that cell cycle progression of cervical cancer cells no longer depends on *miR-15a/miR-16* activity. Since the level of *miR-15a/miR-16* is normally high in these tumours, this may suggest a dual role of *miR-15a/miR-16* as tumour suppressing and oncogenic miRNAs. A similar mechanism may also exist in NSCLC, since both miRNAs were up-regulated in 28% (5/18) of the tumour samples, in three of which Rb was down-regulated.

What are the targets that are responsible for *miR-15a/miR-16*-induced cell cycle arrest in NSCLC? MiRNAs including *miR-15a/miR-16* can affect hundreds of mRNAs (9), which renders it difficult to identify the biologically relevant targets. Although it is believed that *miR-15a/miR-16*-induced cell cycle arrest is mediated by targeting G1 cyclins (10, 11), it cannot be excluded that this is induced by an indirect mechanism, for example by targeting mRNAs from essential genes. Based on our findings, however, that *miR-15a/miR-16*-induced cell cycle arrest depends on Rb, we can conclude that this is due to down-regulation of proteins directly involved in cell cycle control.

It was shown recently that CCND1, CCND3, CCNE1, CDK4 and CDK6 are direct targets of *miR-15a/miR-16* (9-12). We confirm that CCND1, CCND3 and CCNE1 are also targets in NSCLC cell lines and provide evidence that CCND2 is an additional target of these miRNAs. For most of these targets it was unclear, if they are also regulated by physiological concentrations of *miR-15a/miR-16*. We demonstrate that CCND1, CCND2 and CCNE1 are directly regulated by *miR-15a/miR-16* under physiological conditions. Using luciferase reporter constructs containing target sequences from the different cyclin genes we showed that endogenous miRNAs are able to down-regulate luciferase activity and that this effect is reversed by co-transfection with anti-miRs against *miR-15a/miR-16*. *MiR-16* inversely correlates with CCND1 protein in NSCLC, which is in line with the finding that this is a physiological target. In addition, *miR-15a/miR-16*-induced cell cycle arrest can be partially restored by overexpression of CCND1 or CCNE1.

In conclusion, we demonstrate that *miR-15a/miR-16* induce cell cycle arrest by down-regulation of G1 cyclins, while NSCLC cells escape these growth inhibitory signals either by down-regulation or loss of function of Rb, down-regulation of *miR-15a/miR-16* or by other means. This may constitute a general mechanism implicated in tumourigenesis, since LOH 13q14 and Rb inactivation are frequent events in various tumours. Furthermore, our data suggest that *miR-15a/miR-16* might be used as therapeutic agents in Rb-proficient NSCLC.

Acknowledgments

We thank Mario Tschan for plasmids and cell lines, Bernadette Wyder for introduction into flow cytometry and Sabine Jakob and Silvia Rihs for protocols and helpful discussions. This work was supported by a grant from the Swiss National Science Foundation to EV.

References

1. Jemal A, Siegel R, Ward E, Murray T, Xu J, Thun MJ. Cancer statistics, 2007. CA Cancer J Clin 2007;57:43-66.
2. Wistuba II, Gazdar AF. Lung cancer preneoplasia. Annu Rev Pathol 2006;1:331-48.
3. Mattie MD. MicroRNAs in cancer (Oncomirs). In: Clarke NJ and Sanseau P, editors. MicroRNAs: biology, function and expression. Eagleville: DNA Press; 2007. p. 251-80
4. Hwang HW, Mendell JT. MicroRNAs in cell proliferation, cell death, and tumorigenesis. Br J Cancer 2006;94:776-80.
5. Johnson SM, Grosshans H, Shingara J, et al. RAS is regulated by the let-7 microRNA family. Cell 2005;120:635-47.
6. Takamizawa J, Konishi H, Yanagisawa K, et al. Reduced expression of the let-7 microRNAs in human lung cancers in association with shortened postoperative survival. Cancer Res 2004;64:3753-6.
7. Calin GA, Dumitru CD, Shimizu M, et al. Frequent deletions and down-regulation of micro- RNA genes miR15 and miR16 at 13q14 in chronic lymphocytic leukemia. Proc Natl Acad Sci U S A 2002;99:15524-9.
8. Cimmino A, Calin GA, Fabbri M, et al. miR-15 and miR-16 induce apoptosis by targeting BCL2. Proc Natl Acad Sci U S A 2005;102:13944-9.
9. Linsley PS, Schelter J, Burchard J, et al. Transcripts targeted by the microRNA-16 family cooperatively regulate cell cycle progression. Mol Cell Biol 2007;27:2240-52.
10. Chen RW, Bemis LT, Amato CM, et al. Truncation in CCND1 mRNA alters miR-16-1 regulation in mantle cell lymphoma. Blood 2008;112:822-9.
11. Liu Q, Fu H, Sun F, et al. miR-16 family induces cell cycle arrest by regulating multiple cell cycle genes. Nucleic Acids Res 2008;36:5391-404.
12. Bonci D, Coppola V, Musumeci M, et al. The miR-15a-miR-16-1 cluster controls prostate cancer by targeting multiple oncogenic activities. Nat Med. 2008;14:1271-7.

13. Hinds P, Mittnacht S, Dulic V, Arnold A, Reed S, Weinberg R. Regulation of retinoblastoma protein functions by ectopic expression of human cyclins. Cell 1992;70:993-1007.

14. Tan X, Martin SJ, Green DR, Wang JYJ. Degradation of Retinoblastoma protein in tumor necrosis factor- and CD95-induced cell death. J Biol Chem 1997;272:9613-6.

15. Gugger M, Kappleler A, Vonlanthen S, et al. Alterations of cell cycle regulators are less frequent in advanced non-small cell lung cancer than in resectable tumours. Lung Cancer 2001;33:229-39.

16. Campbell I. Chi-squared and Fisher-Irwin tests of two-by-two tables with small sample recommendations. Stat Med 2007;26:3661-75.

17. Masuda N, Ohnishi T, Kawamoto S, Monden M, Okubo K. Analysis of chemical modification of RNA from formalin-fixed samples and optimization of molecular biology applications for such samples. Nucleic Acids Res 1999;27:4436-43.

18. Chivukula RR, Mendell JT. Circular reasoning: microRNAs and cell-cycle control. Trends Biochem Sci 2008.

19. Lukas J, Bartkova J, Rohde M, Strauss M, Bartek J. Cyclin D1 is dispensable for G1 control in retinoblastoma gene-deficient cells independently of cdk4 activity. Mol Cell Biol 1995;15:2600-11.

20. McGahren-Murray M, Nicholas H, Keyomarsi K. The differential suaurosporine-mediated G1 arrest in normal versus tumor cells is dependent on the retinoblastoma protein. Cancer Res 2006;66:9744-53.

21. Evan GI, Vousden KH. Proliferation, cell cycle and apoptosis in cancer. Nature 2001;411:342-8.

22. Karube Y, Tanaka H, Osada H, et al. Reduced expression of Dicer associated with poor prognosis in lung cancer patients. Cancer Sci 2005;96:111-5.

23. Chung EY, Dews M, Cozma D, et al. c-Myb oncoprotein is an essential target of the dleu2 tumor suppressor microRNA cluster. Cancer Biol Ther 2008;7.

24. Calin GA, Ferracin M, Cimmino A, et al. A MicroRNA signature associated with prognosis and progression in chronic lymphocytic leukemia. N Engl J Med 2005;353:1793-801.

25. Leversha MA, Fielding P, Watson S, Gosney JR, Field JK. Expression of p53, pRB, and p16 in lung tumours: a validation study on tissue microarrays. J Pathol 2003;200:610-9.

26. Bruce J, Hurford RJ, Classon M, Koh J, Dyson N. Requirements for cell cycle arrest by p16INK4a. Mol Cell 2000;6:737-42.

27. Modi S, Kubo A, Oie H, Coxon AB, Rehmatulla A, Kaye FJ. Protein expression of the RB-related gene family and SV40 large T antigen in mesothelioma and lung cancer. Oncogene 2000;19:4632-9.

28. Stilgenbauer S, Nickolenko J, Wilhelm J, et al. Expressed sequences as candidates for a novel tumor suppressor gene at band 13q14 in B-cell chronic lymphocytic leukemia and mantle cell lymphoma. Oncogene 1998;16:1891-7.

29. Elnenaei MO, Hamoudi RA, Swansbury J, et al. Delineation of the minimal region of loss at 13q14 in multiple myeloma. Genes Chrom. Cancer 2003;36:99-106.

30. Bieche I, Lidereau R. Loss of heterozygosity at 13q14 correlates with RB1 gene underexpression in human breast cancer. Mol Carcinog 2000;29:151-8.

31. Dong JT, Boyd JC, Frierson HF Jr. Loss of heterozygosity at 13q14 and 13q21 in high grade, high stage prostate cancer. Prostate 2001;49:166-71.

32. Bottoni A, Piccin D, Tagliati F, Luchin A, Zatelli MC, degli Uberti EC. miR-15a and miR-16-1 down-regulation in pituitary adenomas. J Cell Physiol 2005;204:280-5.

33. Wang X, Tang S, Le SY, Lu R, Rader JS, Meyers C, Zheng ZM. Aberrant expression of oncogenic and tumor-suppressive microRNAs in cervical cancer is required for cancer cell growth. PLoS ONE 2008;3:e2557.

Manuscript 1

Supplementary figures

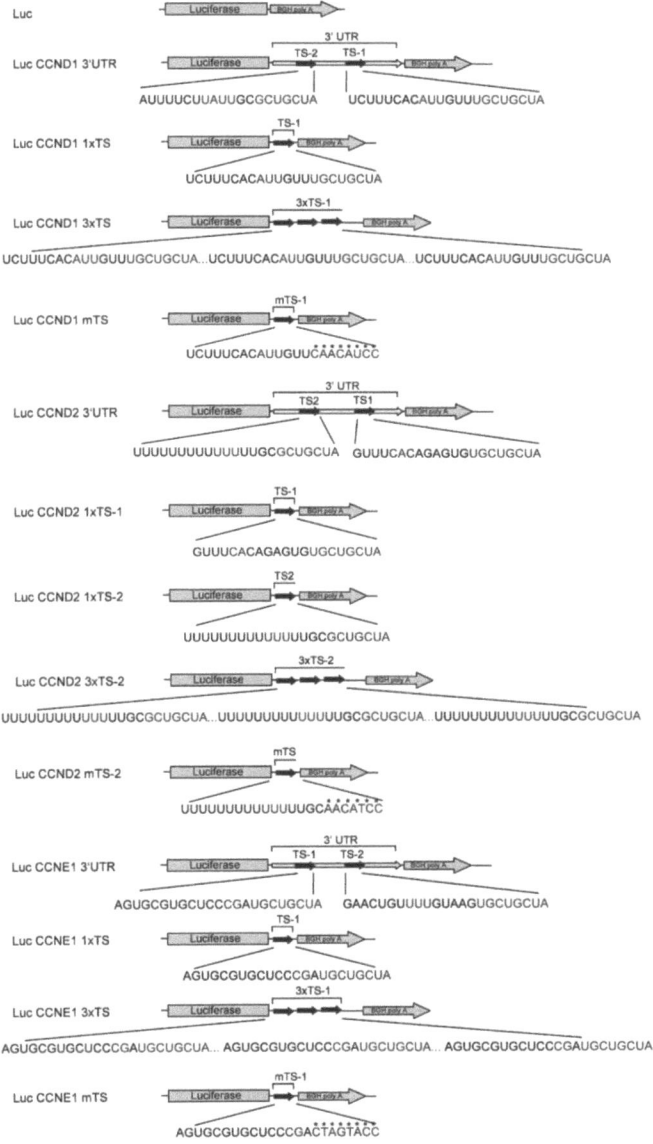

Supplementary Fig. 1. Schematic depiction of the parental Luc construct and progenitor luciferase constructs containing *miR-15a* and *miR-16* target sites (TS) from *CCND1*, *CCND2* and *CCNE1*, respectively. Sequences of the different TS are indicated. Mutated sequences are indicated by asterisks.

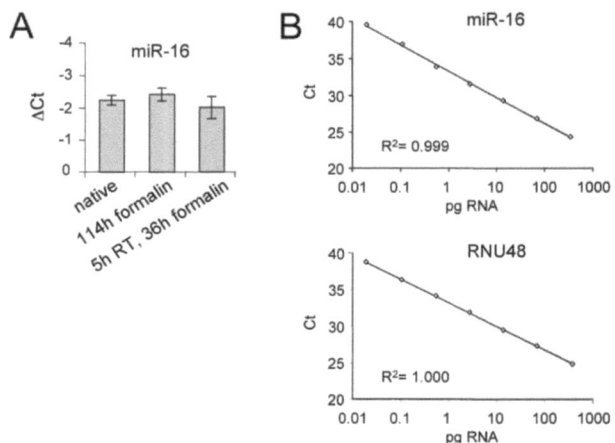

Supplementary Fig. 2. Quantification of *miR-16* from formalin-fixed paraffin-embedded tissues. (A) Quantification of *miR-16* relative to RNU48 in native lung tissue or lung tissues fixed for 114 hrs or 36 hrs in formalin, respectively; the latter was left at room temperature for 5 hrs prior to fixation. (B) Serial dilutions of cDNAs from A549 reveal that *miR-16* quantification is linear over a wide concentration range.

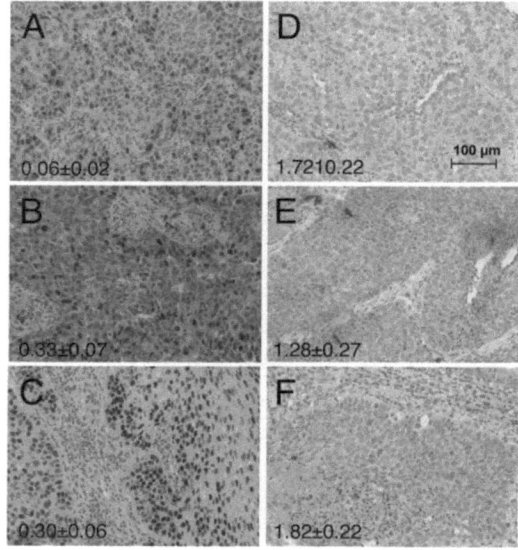

Supplementary Fig. 3. Inverse correlation of *miR-16* and CCND1 in NSCLC. (A, B, C) Examples of CCND1-positive tumours as indicated by immunohistochemistry. (D, E, F) Examples of CCND1-negative tumours. Signal intensity was relative to IgG1. Same tumour areas on serial tissue sections were analysed. Numbers indicate the level *miR-16* expression in the tumour tissue relative to corresponding normal tissue from the same paraffin block.

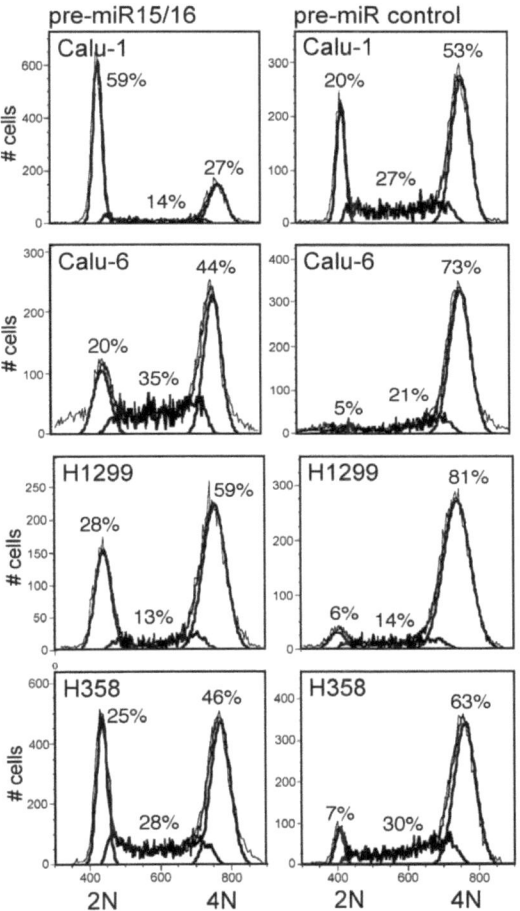

Supplementary Fig. 4. Cell cycle analysis of propidium iodide-stained NSCLC cells. Cells were co-transfected with *miR-15a* and 16 precursors and treated with 100 ng/ml nocodazole for 20 hrs beginning at 24 hrs post transfection. Cells were stained with propidium iodide and the DNA content of individual cells was quantified by flow cytometry. 2N, cells having diploid DNA content; 4N, cells having tetraploid DNA content.

```
CCND1 3'UTR position 1961-1967   5'...AUUUUCUUAUUGC---GCUGCUA...
                                         ||||        |||||||
miR-16                           3'   GCGGUUAUAAAUGCACGACGAU

CCND1 3'UTR position 2033-2040   5'...UCUUUCACAUUGUU-UGCUGCUA...
                                          |||         ||||||||
miR-16                           3'   GCGGUUAUAAAUGCACGACGAU

CCND2 3'UTR position  620-626    5'...UGCUGUCUACAGUA--GCUGCUA...
                                           ||          |||||||
miR-16                           3'   GCGGUUAUAAAUGCACGACGAU

CCND2 3'UTR position 1693-1699   5'...UUUUUUUUUUUUUUGCGCUGCUA...
                                            |||       |||||||
miR-16                           3'   GCGGUUAUAAAUGCACGACGAU

CCND2 3'UTR position 1764-1771   5'...GUUUCACAGAGUG--UGCUGCUA...
                                           ||         ||||||||
miR-16                           3'   GCGGUUAUAAAUGCACGACGAU

CCND3 3'UTR position  760-774    5'...UGGCCAAGGA------GCUGCUA...
                                         |||||        |||||||
miR-16                           3'   GCGGUUAUAAAUGCACGACGAU

CCNE1 3'UTR position  247-254    5'...AGUGCGUGCUCCCGAUGCUGCUA...
                                                    || ||||||||
miR-16                           3'   GCGGUUAUAAAUGC-ACGACGAU

CCNE1 3'UTR position  485-492    5'...GAACUGUUUUGUAAGUGCUGCUA...
                                              |||      ||||||||
miR-16                           3'   GCGGUUAUAAAUGC--ACGACGAU
```

Supplementary Fig. 5. Predicted target sites of *miR-15a* and 16 within the 3' untranslated region of G_1 cyclins. Both miRNAs are predicted to bind to the same regions, but for reasons of simplicity, only the sequence alignment to *miR-16* is shown. Predictions are from TargetScan with the exception of *CCND3*. Only sites that are conserved in human, mouse, rat and dog are shown.

Manuscript 1

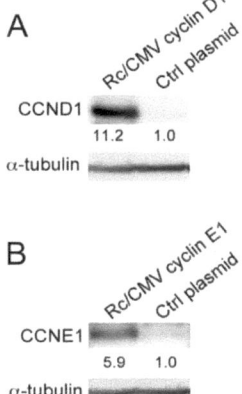

Supplementary Fig. 6. Western blot analysis of CCND1 and CCNE1 over-expressors. A549 cells were co-transfected with Rc/CMV cyclin D1, Rc/CMV cyclin EΔ3'UTR or control plasmid in combination with *miR-15a/miR-16* precursors and analysed by Western blot 48 hrs post transfection using monoclonal antibodies as indicated in the figure. Protein levels were normalized to α-tubulin and presented relative to the control.

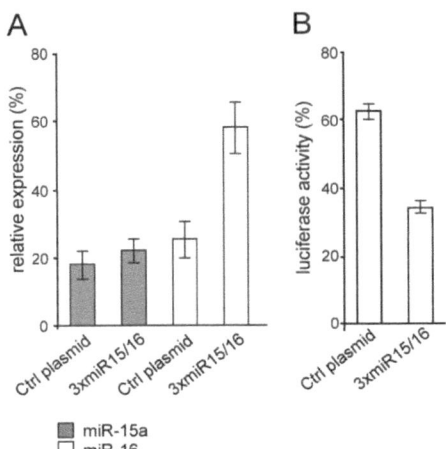

Supplementary Fig. 7. *MiR-15a* and *miR-16* down-regulation affects target gene expression in a NSCLC cell line. H2009 was co-transfected with pcDNA3 3xmiR15/16 (containing three copies of the *miR-15a/miR-16* locus cloned in tandem into pcDNA3) or pcDNA3 control in combination with luciferase constructs. Cells were harvested 48 hrs post transfection and analysed for microRNA expression and luciferase activity. (A) Steady state level of *miR-15a* and *miR-16*. MiRNA levels were quantified relative to the level obtained in normal lung tissue (see Fig. 1B). (B) Luciferase activity obtained with the construct Luc CCND1 3xTS relative to the activity obtained with the Luc construct containing no target site.

Manuscript 2

miR-34a and *miR-15a/16* are co-regulated in non-small cell lung cancer and control cell cycle progression in a synergistic and Rb-dependent manner

Nora Bandi and Erik Vassella*

Molecular Cancer 2011; 10:55.

Institute of Pathology, University of Bern, Bern, Switzerland

*Corresponding author: E. Vassella
Institute of Pathology
University of Bern
Murtenstrasse 31
CH-3010 Bern
Switzerland

phone: 0041-31-632 9943
fax: 0041-31-381 8764
Email: erik.vassella@pathology.unibe.ch

Running title: cell cycle control by microRNAs in lung cancer
Key words: microRNA, synergism, non-small cell lung cancer, cell cycle control, retinoblastoma

miR-34a and *miR-15a/16* are co-regulated in non-small cell lung cancer and control cell cycle progression in a synergistic and Rb-dependent manner

Nora Bandi and Erik Vassella

Institute of Pathology, University of Bern, Bern, Switzerland

Abstract

Background: MicroRNAs (miRNAs) are small non-coding RNAs that are frequently involved in carcinogenesis. Although many miRNAs form part of integrated networks, little information is available how they interact with each other to control cellular processes. *miR-34a* and *miR-15a/16* are functionally related; they share common targets and control similar processes including G1-S cell cycle progression and apoptosis. The aim of this study was to investigate the combined action of *miR-34a* and *miR-15a/16* in non-small cell lung cancer (NSCLC) cells.

Methods: NSCLC cells were transfected with *miR-34a* and *miR-15a/16* mimics and analysed for cell cycle arrest and apoptosis by flow cytometry. Expression of retinoblastoma and cyclin E1 was modulated to investigate the role of these proteins in miRNA-induced cell cycle arrest. Expression of miRNA targets was assessed by real-time PCR. To investigate if both miRNAs are co-regulated in NSCLC cells, tumour tissue and matched normal lung tissue from 23 patients were collected by laser capture microdissection and compared for the expression of these miRNAs by real-time PCR.

Results: In the present study, we demonstrate that *miR-34a* and *miR-15a/16* act synergistically to induce cell cycle arrest in a Rb-dependent manner. In contrast, no synergistic effect of these miRNAs was observed for apoptosis. The synergistic action on cell cycle arrest was not due to a more efficient down-regulation of targets common to both miRNAs. However, the synergistic effect was abrogated in cells in which cyclin E1, a target unique to *miR-15a/16*, was silenced by RNA interference. Thus, the synergistic effect was due to the fact that in concerted action both miRNAs are able to down-regulate more targets involved in cell cycle control than each miRNA alone. Both miRNAs were significantly co-regulated in adenocarcinomas of the lung suggesting a functional link between these miRNAs.

Conclusions: In concerted action miRNAs are able to potentiate their impact on G_1-S progression. Thus the combination of miRNAs of the same network rather than individual miRNAs should be considered for assessing a biological response. Since *miR-34a* and *miR-15a/16* are frequently down-regulated in the same tumour tissue, administrating a combination of both miRNAs may also potentiate their therapeutic impact.

Background

Lung cancer is the leading cause of cancer-related death in industrialized countries [1]. Systemic treatment of lung cancer patients includes chemotherapy, inhibitors of angiogenesis and inhibitors of EGFR signaling. However, since the effect of these drugs is only transient, the overall five-year survival rate is less than 15%. Non-small cell lung carcinoma (NSCLC) accounts for 80% of lung cancer and is further subdivided into two major types, squamous cell carcionoma and adenocarcimoma [2]. Squamous cell carcinoma usually arises from the major bronchi, whereas adenocarcinoma arises from distant airway bronchioles and alveoli. These tumours show frequent alterations of genes involved in cell cycle control or apoptosis including *k-RAS*, *EGFR*, *c-Myc*, *cyclin D1* (*CCND1*), *TP53*, *retinoblastoma* (*Rb*), *p16INK* and *Bcl2* [3], but the relevant molecular mechanisms driving the aggressive biological behaviour of these tumours are largely unknown.

miRNAs are small regulatory RNA molecules at the post-transcriptional level and are implicated in a wide variety of biological processes including proliferation, differentiation and apoptosis [4]. Notably, miRNAs form networks to regulate the expression of individual components of the cell cycle control machinery. Many of these miRNAs including the *let-7* family [5], *miR-34* [6], *miR-15a/16* [7], *miR-221/222* [8, 9], *miR-17-92* [10], *miR-107* and *miR-185* [11] are frequently dysregulated in lung cancer and therefore constitute promising targets for specific anticancer intervention (reviewed by Negrini et al. [12]).

Many miRNAs are implicated in cell cycle progression or apoptosis, but surprisingly little information is available if these miRNAs are able to interact with each other to co-ordinately regulate these cellular processes. In addition, it is poorly understood why miRNAs often share common targets despite the fact that they constitute a relatively small family of RNAs encoded by less than 1000 genes. In this study we have analysed two miRNAs, *miR-15a/16* and *miR-34*, which are located at chromosomal regions 13q14 and 1p34, respectively. Although these miRNAs contain completely unrelated seed sequences, they are functionally related since they are both able to induce G_1-G_0 cell cycle arrest and apoptosis [7, 13-15]. In addition, they share common targets including *CCND1*, *CDK4*, *CDK6*, *E2F3* and *Bcl2*. However, other targets also exist which are unique to *miR-15a/16* (*cyclin E1 (CCNE1)*, *cyclin D2 (CCND2)* or *cyclin D3 (CCND3)*) or *miR-34a* (*c-Myc, n-Myc,* and *c-Met*) [7, 16-18].

To investigate if these miRNAs are able to interact with each other for the regulation of cellular processes, they were overexpressed in NSCLC cell lines. Here we demonstrate that *miR-15a/16* and *miR-34* act synergistically to induce cell cycle arrest in a Rb-dependent manner. The synergistic effect can be explained by the fact that in concerted action, miRNAs are able to down-regulate more targets than each miRNA alone. Thus, it may be important to analyse miRNAs in a combinatorial mode as this may provide additional information on their role in specific cellular processes. The significance of the growth-inhibitory effect by the combined action of these miRNAs is further supported by the finding that they are co-regulated in adenocarcinomas and squamous cell carcinomas of the lung. Our results suggest that targeting a combination of miRNAs involved in the same pathway may potentiate the therapeutic effect of each individual miRNA.

Materials and Methods

Cell lines and culture conditions

The NSCLC cell lines A549, H2009, H1299 and H358 were obtained from the American Type Culture Collection, Rockville, MD. All cell lines were cultured in Iscove's modified Dulbecco's medium supplemented with 2 mM L-alanyl-L-glutamine, 1% penicillin/streptomycin and 5% foetal bovine serum (Sigma) at 37°C and 5% CO_2.

Transfection

Cells were seeded in culture flasks 24 h prior to transfection. Co-transfections with plasmid DNA were performed using Effectene reagent (Qiagen), all other transfections were performed using HiPerFect (Qiagen). If not otherwise specified, transfection was performed using 20 nM of hsa-*pre-miR-34a*, a mixture of 10 nM hsa-*pre-miR-15a* and 10 nM hsa-*pre-miR-16* or 20 nM pre-miR miRNA precursor control 1 (Ambion). Si RNAs against Rb, or CCNE1 (siGENOME SMARTpool, Dharmacon) were used at 60 nM or 7.8 nM, respectively. Control transfections were performed using non-targeting Pool 2 (siGENOME). pCMV-Rb [19] or empty control plasmid were used at 125 ng/ml. Transfection efficiency of short RNAs and plasmid DNA was monitored using siGloGreen transfection indicator (Dharmacon) or an RFP-expression plasmid, respectively.

Cell cycle analysis and cell death assay

Cell cycle analysis was performed by flow cytometry essentially as described [7]. For cell death analysis, floating and adherent cells were harvested, combined, washed with PBS and stained with 10 ug/ml propidium idode (Sigma). Apoptotic cells were detected using an antibody directed against cleaved caspase 3 (clone 5A1E, 1:100, Cell Signaling) as described [20]. Cells were analysed using an LSR II flow cytometer (BD Biosciences) and FlowJo 8.8.4 software (Tree Star). As a positive control, cells were treated with UV (400mJ) using a UV Stratalinker 1800 (Stratagene).

RNA isolation and real-time PCR

Total RNA was extracted from cultured cells using the miRVana RNA isolation kit according to the manufacturer's instructions (Ambion). TaqMan miRNA assays (Applied Biosystems) were performed as described [7] using a Real-Time PCR system 7500 (Applied Biosystems). miRNA levels were normalized to the level obtained for *RNU48*. Quantification of *Bcl2* was done using a TaqMan assay (Applied Biosystems); all other mRNAs were quantified using Quantitec primer assays (Qiagen). mRNA levels were normalized to the level obtained for *GAPDH*. Changes in expression were calculated using the $\Delta\Delta Ct$ method.

Western blot analysis

Western blot analysis was performed as described [7]. Monoclonal antibodies against Rb (clone 3C8, QED Bioscience) and phospho-Rb (Ser807/811, Cell Signaling Technology) were diluted 1:1000, monoclonal antibody against Bcl2 (clone 124, Dako) was diluted 1:300, and monoclonal antibody against α-tubulin (clone B512, Sigma) was diluted 1:5000. Secondary goat anti-mouse-HRP and goat anti-rabbit-HRP antibodies (Biorad) were used at 1:5000 or 1:7000, respectively.

Laser capture microdissection

Formalin-fixed paraffin-embedded tissues from 11 adenocarcinomas and 12 squamous cell carcinomas were used for miRNA expression analysis. Tumour tissues and corresponding normal tissues from bronchiolar or alveolar epithelium, respectively, was collected by laser capture microdissection as described previously [7]. Stroma components including connective tissues, inflammatory cells and blood vessels were excluded. Nucleic acids were subjected to a heat-treatment in order to remove methylol groups introduced during formalin fixation and subjected to real-time PCR as described [7]. All experiments using human specimens were done according to the ethical guidelines of the Institute of Pathology, University of Bern, and were reviewed by the institutional review board.

Statistics

Statistical analyses were performed using the GraphPAD prism software. Statistical differences were calculated using unpaired two-tailed student's t-test. A probability of $p \leq 0.05$ was considered statistically significant. Statistical significance of correlation was assessed by the Pearson test.

Results

miR-15a/16 and *miR-34a* are co-regulated in adenocarcinomas of the lung

The finding that *miR-15a/16* and *miR-34a* share many common targets involved in G_1 progression prompted us to analyse if these miRNAs are co-regulated in NSCLC. Tumour tissues of adenocarcinomas and squamous cell carcinomas and matched normal tissues from alveolar or bronchial epithelium, respectively, were collected by laser capture microdissection and compared for the expression of both miRNAs by real-time PCR. We have shown previously that *miR-15a and miR-16* are frequently down-regulated in NSCLC [7]. Here we demonstrate that the expression of *miR-16* is significantly correlated to the expression of *miR-34a* in adenocarcinomas (p=0.018), but not in squamous cell carcinomas of the lung. Both miRNAs were down-regulated in 82% (9/11), and upregulated in 18% (2/11) of adenocarcinomas (Fig. 1A). In contrast, *miR-34a* was significantly down-regulated in 10/12 squamous cell carcinoma samples while *miR-16* was down-regulated in 5/12 tumour samples. Both miRNAs were also significantly down-regulated in the NSCLC cell lines A549, H2009, H1299 and H358 (data not shown).

Co-repression of miRNAs may be due to defects in miRNA processing. Notably, reduced expression of Dicer has been detected in NSCLC [21]. To address this possibility, the same tumour tissues were analysed for the expression of *miR-21*, which is frequently upregulated in lung cancer [22, 23]. As shown in Fig. 1B, *miR-21* was upregulated or expressed at normal levels in 9 of 11 adenocarcinomas and 7 of 12 squamous cell carcinomas. Thus, abrogation of miRNA processing may only account for a small subgroup of NSCLC samples.

We next investigated the possibility that both miRNAs are linked because they are able to mutually regulate their expression. To this end, *miR-15a/16* or *miR-34a* were overexpressed by transfection with miRNA precursors (pre-miRNA) in a Rb-deficient NSCLC cell line, H2009, and analysed for the expression of the miRNA counterpart. Since H2009 is refractory to cell cycle arrest induced by *miR-15a/16* [7] and *miR-34a* (see below), secondary effects on miRNA expression as a consequence of the G_1-G_0 arrest can be excluded using this cell line. However, neither *miR-15a/16* nor *miR-34a* was able to affect the expression of its counterpart, while *CDK6* mRNA, a target common to both miRNAs, was significantly down-regulated (Fig. 1C).

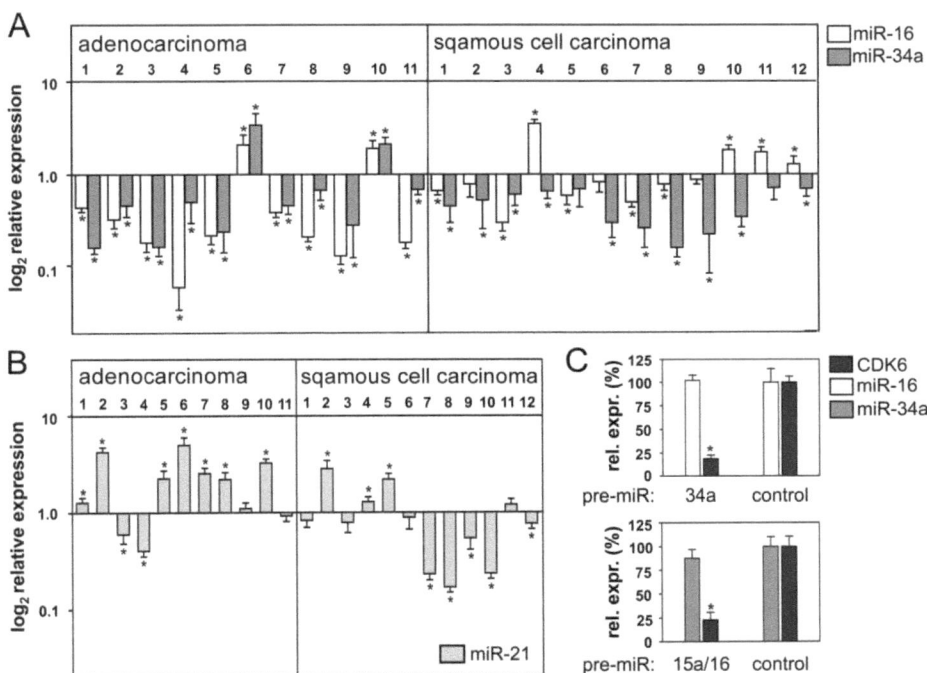

Figure 1. *miR-34a* and *miR-15a/16* are co-regulated in non-small cell lung cancer. (A) miR-16 and *miR-34a* levels in adenocarcinomas and squamous cell carcinomas relative to matched normal tissues. (B) *miR-21* levels in the same tumour samples, relative to matched normal tissue. (C) *miR-34a* and *miR-15a/16* are not able to mutually regulate their expression H2009 cells were transfected with *pre-miR-34a* (upper panel) or *pre-miR-15a/16* (lower panel) and analysed for the expression of the miRNA counterpart or *CDK6* mRNA 42 h post-transfection. Values are relative to the value obtained for the control transfected with precursor control (n=3). *, P < 0.05.

miR-34a-induced cell cycle arrest depends on the expression of Rb

One miRNA can affect the expression of hundreds of proteins [24], which often renders it difficult to identify the relevant targets. We have previously shown that *miR-15a/16*-induced cell cycle arrest depends on the expression of Rb. This indicates that components of the cell cycle machinery upstream of Rb, including *cyclin D1* (*CCND1*), *cyclin D3* (*CCND3*), *CDK4* and *CDK6* [7, 25, 26], are the most relevant targets of miR-15a/16. To investigate if miR15a/16 and miR-34a share redundant functions, miR-34a was analysed using the same set of experiments as we have described previously for miR-15a/16 [7].

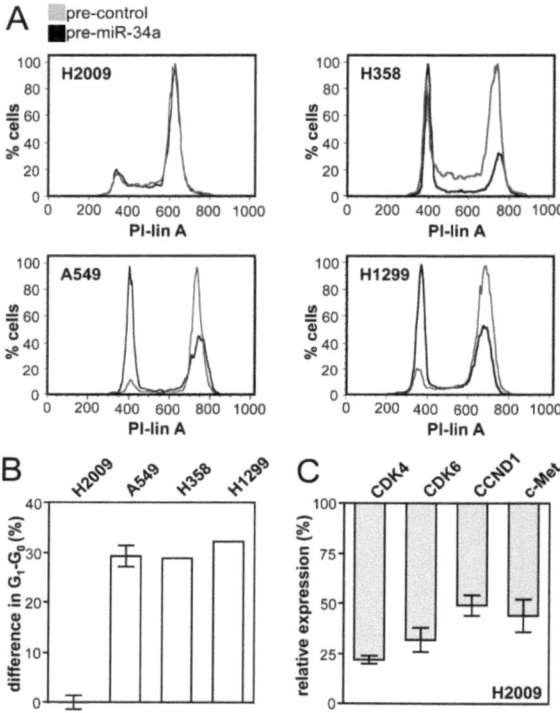

Figure 2. H2009 cells are refractory to miR-34a-induced cell cycle arrest. (A) DNA content distribution of NSCLC cells transfected with pre-miR precursors or precursor control. Cells were treated for 18 h with nocodazole beginning 24 h post-transfection. (B) Percent difference in G_1-G_0 between cells transfected with pre-miR-34a and cells transfected with precursor control. H2009, A549, n=3; H1299 and H358, n=1. (C) mRNA levels of known miR-34a targets. H2009 cells were transfected with pre-miR-34a and harvested 42 h post-transfection (n=3). Values are relative to the level obtained for the control transfected with precursor control.

To investigate cell cycle arrest, the NSCLC cell lines A549, H358, H1299 and H2009 were transfected with pre-miRNA-34a and treated with nocodazole 24 h post-transfection. Nocodazole traps cells at the G_2-M phase, but ~30% of the transfected A549, H358 or H1299 cells accumulated in G_1-G_0 (Fig. 2A and 2B) indicating that miR-34a induces an arrest in this phase of the cell cycle. In contrast, Rb-deficient H2009 cells were completely refractory to miR-34a-induced arrest (Fig. 2A and 2B). However, known targets of miR-34a including *CDK4*, *CDK6*, *CCND1* and *c-Met* were significantly down-regulated in these cells (Fig. 2C).

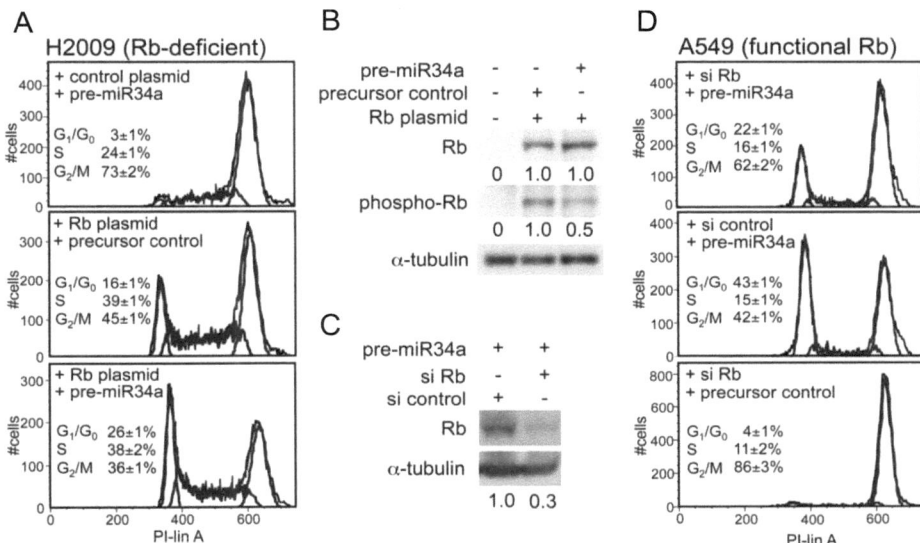

Figure 3. *miR-34a*-**induced cell cycle arrest depends on the expression of Rb.** (A, D), DNA content distribution by flow cytometry of H2009 (A) and A549 cells (D) treated with nocodazole beginning 48 h post-transfection (n=3). (B, C), Western blot analysis of H2009 (B) and A549 cells (C) subjected to the same conditions as in (A) and (D) using antibodies directed against Rb or phospho-Rb. Protein levels were normalised to α-tubulin.

Rb reconstitution into Rb-deficient NSCLC lines restores G_1 arrest mechanisms [27]. To investigate if *miR-34a*-induced cell cycle arrest depends on the expression of Rb, the latter gene was reintroduced into H2009 cells. Co-transfection of H2009 cells with *pre-miR-34a* and an empty control plasmid induced cell cycle arrest in 3.1±1% of the population (Fig. 3A). Consistent with previous findings [27], the percentage of cells in the G_1-G_0 phase of the cell cycle increased to 16±1% upon transfection with a Rb expression plasmid (p=0.001). This indicates that Rb per se is able to induce cell cycle arrest in a significant proportion of the population. However, Rb plasmid in combination with *pre-miR-34a* induced cell cycle arrest in significantly more cells (26±1%, p=0.01) than Rb plasmid in combination with the precursor control. The expression of Rb protein was not affected by co-transfection with *pre-miR-34a*. In contrast, phospho-Rb was reduced by 50% (Fig. 3B). In conclusion, the ability of *miR-34a* to induce cell cycle arrest depends on the expression of Rb. Complementary experiments were performed in A549 cells, in which the Rb gene was knocked down by RNA interference. The knock-down expressed three times less Rb protein than the control (Fig. 3C). As expected, the knock-down was significantly more resistant to *miR-34a*-induced cell cycle arrest (22±1% in G_1-G_0) than the control (43±1% in G_1-G_0, p<0.001) (Fig. 3D). In conclusion, there is a significant degree of redundancy between *miR-34a* and *miR-15a/16* in their ability to induce cell cycle arrest in NSCLC cells.

miR-15a/16 and *miR-34a* act synergistically to induce arrest in G_1-G_0

We next addressed the question if *miR-15a/16* and *miR-34a* act together to induce cell cycle arrest. A549 cells were transfected with increasing concentrations of *pre-miR-15a/16* or *pre-miR-34a* and analysed for cell cycle arrest (Fig. 4A).

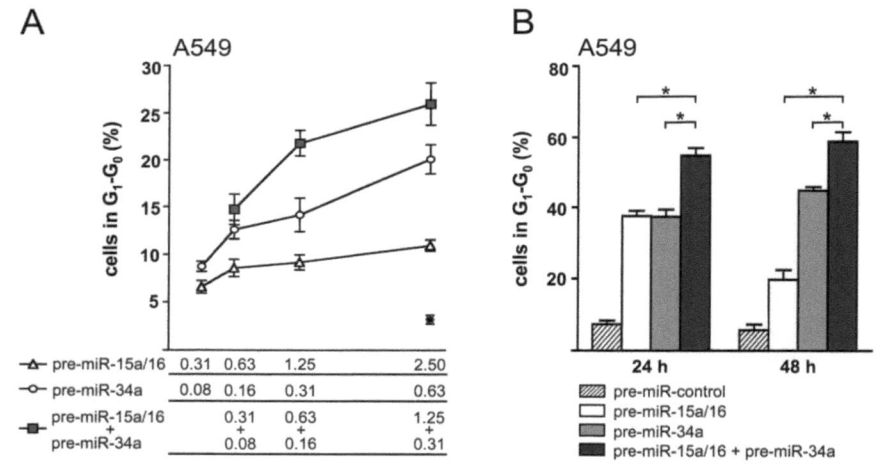

Figure 4. *miR-15a/16* and *miR-34a* act synergistically to induce cell cycle arrest. (A) Cell cycle analysis of A549 cells transfected with *pre-miR-34a* and/or *pre-miR-15a/16* under non-saturating conditions. Precursors were supplemented with precursor control to yield a total concentration of 2.5 nM per transfection. *, transfection with 2.5 nM precursor control. Cells were treated for 18 h with nocodazole beginning 24 h post-transfection (n=3). (B) Cell cycle analysis under saturating conditions. A549 cells were transfected with 20 nM precursor or precursor control or co-transfected with 10 nM *pre-miR-34a* and 10 nM *pre-miR-15a/16* and treated for 18 h with nocodazole beginning 24 h (left panel) or 48 h (right panel) post-transfection (n=3).

From the slope of the dose-response curves it can be deduced that *pre-miR-34a* was more efficient than pre-*miR-15a/16* in inducing cell cycle arrest.

We next assessed the concerted action of *miR-15a/16* and *miR-34a* precursors on cell cycle arrest. Transfection with 2.5 nM *pre-miR-15/16* or transfection with 0.63 nM *pre-miR-34a* resulted in a G_1-G_0 arrest of A549 cells in 20.1±1.6% and 10.9±0.6% of the population, respectively.

Interestingly, a mixture with half the concentrations of *pre-miR-15/16* and *pre-miR-34a* was more efficient (25.9±2.2%) than each pre-miRNA alone in inducing a G_1-G_0 arrest, $p \leq 0.02$ (Fig. 4A). Consistent results were obtained over a four-fold concentration range. Thus, these results clearly indicate that *miR-15a/16* and *miR-34a* act synergistically to induce cell cycle arrest in G_1-G_0.

Both *pre-miRNAs* displayed saturation for cell cycle arrest at 20 nM (data not shown). Interestingly, a synergistic effect was also obtained at this concentration: cells transfected with 20 nM *pre-miR-15a/16* or 20 nM *pre-miR-34a* in each case gave rise to a G_1-G_0 arrest in about 37% of the population. In contrast, co-transfection of cells with both pre-miRNAs at half the concentration (10 nM each) resulted in a G_1-G_0 arrest in 54.6±2.2% of the population (Fig. 4B, 24 h), $p<0.0005$. The synergistic effect was observed 24 h and 48 h post-transfection (Fig. 4B).

No synergistic action on cell death

miR-15a/16 and *miR-34a* are both able to down-regulate the anti-apoptotic protein *Bcl2* (Fig. 5A). In addition, both miRNAs have been shown to induce apopotosis in different cell systems [13, 15].

We therefore wondered if these miRNAs also act synergistically to induce apoptosis. In line with previous findings we were unable to detect any significant increase in propidium iodide (PI)-positive A549 and H2009 cells on transfection with *miR-15a/16* (Fig. 5B and Suppl. Fig. S1).

Likewise, no significant increase in cleaved caspase-3-positive cells was observed (Fig. 5C). In contrast, *miR-34a* elicited a 2-3-fold increase in PI-positive cells and an eight-fold increase in cleaved caspase-3-positive cells 72-96 h post-transfection (Fig. 5B and 5C). Surprisingly, a mixture of both pre-miRNAs at half the concentration was less efficient than *miR-34a* alone in inducing cell death. In conclusion, no synergism elicited by *miR-15a/16* and *miR-34a* exists for cell death in NSCLC cells.

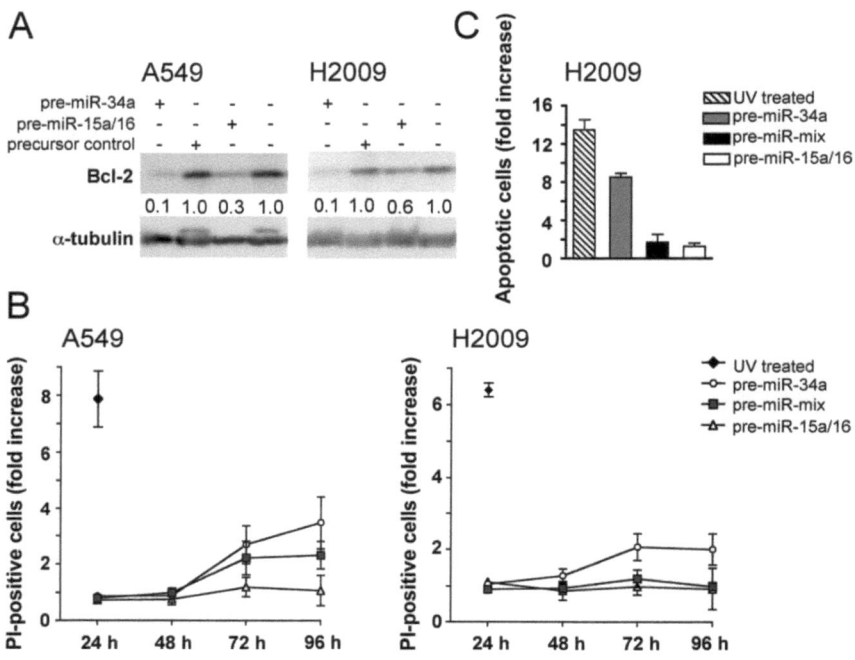

Figure 5. No synergism on cell death. (A) Bcl2 expression. A549 or H2009 cells were transfected with 20 nM precursors and harvested 42 h post-transfection. Western blot analysis was performed using a monoclonal antibody against Bcl2. Protein levels were normalized to α-tubulin and presented relative to the level obtained for the control. (B) Time-course of propidium iodide (PI)-positive A549 and H2009 cells by flow cytometry. Cells were transfected with concentrations of precursors as indicated in the legend to Fig. 4B (n=3). Cells were gated as shown in suppl. figure S1A. (C) Cleaved caspase-3-positive cells. Cells were analysed for the presence of cleaved caspase 3 by flow cytometry 72 h post-transfection (n=3). Values are relative to the level obtained for the control transfected with precursor control. As a positive control, cells were treated with UV.

Combined action of *miR-15a/16* and *miR-34a* on the stability of individual mRNA targets

We next investigated the mechanism underlying the synergistic action of miR-34a and miR-15a/16. It is possible that the observed effect is due to a more efficient down-regulation of targets common to both miRNAs by a combined action of miR-34a and miR-15a/16. To circumvent adverse secondary effects on target gene expression of individual G1 proteins as a consequence of cell cycle arrest, the cell line H2009 was again used for the experiment.

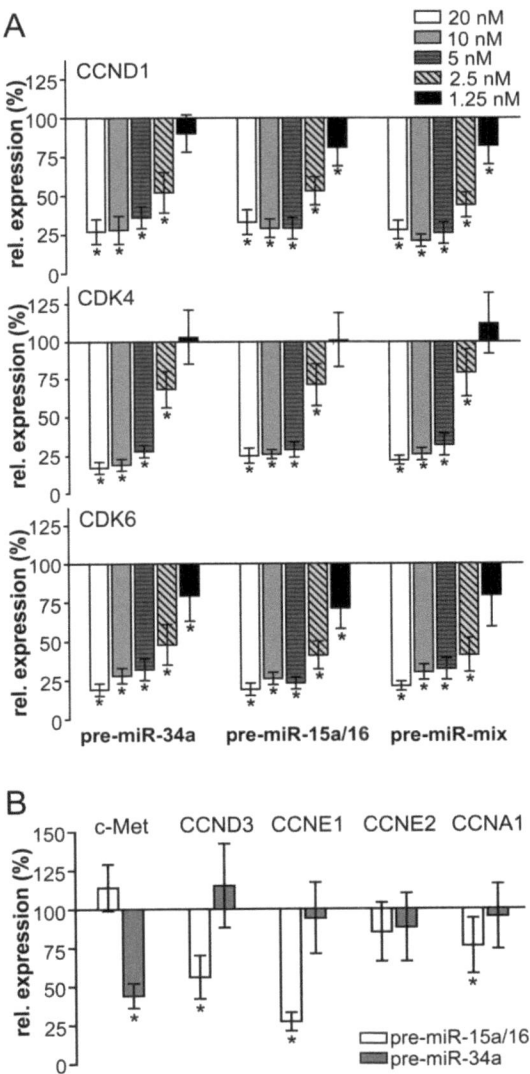

Figure 6. Concerted action of *miR-15a/16* and *miR-34a* on individual mRNA targets. (A) mRNA levels of targets common to both miRNAs. H2009 cells were transfected with pre-miR-15a/16 or pre-miR-34a alone or co-transfected with both pre-miRNAs together (pre-miR-mix) at concentrations as indicated in the figure. (B) Expression level of targets unique to *miR-15a/16* or *miR-34a*. Analysis was performed as described in the legend to Fig. 4C (n=3). *, $p < 0.05$.

Transfection with serial dilutions of pre-miRNAs allowed the establishment of a direct relationship between the amount of input pre-miRNA and the steady-state level of mRNA of the target genes CDK4, CDK6 and CCND1, respectively (Fig. 6A and suppl. Fig. S2).

Interestingly, transfection of H2009 cells with miR-34a or miR-15a/16, or co-transfection of H2009 cells with a mixture of both pre-miRNAs at half the concentration gave rise to comparable dose-response curves for all three target genes (Fig. 6A). From these results we may conclude that miR-15a/16 and miR-34a act in an additive rather than in a synergistic manner on individual mRNAs. Comparable results were obtained in H2009 and A549 cells excluding the possibility that the concerted action on individual mRNA targets depends on the expression of Rb (data not shown).

The synergistic action of *miR-15a/16* and *miR-34a* is due to the down-regulation of additional genes involved in G_1 progression

Although *miR-15a/16* and *miR-34a* share many common targets, other targets exist which are unique to *miR-15a/16* or *miR-34a* (suppl. Fig. S2). Since no synergistic effect was observed for individual mRNA targets, we hypothesized that the observed effect may be due to the fact that more targets involved in G_1–S progression are repressed by the combined action of both miRNAs. Target specificity was re-evaluated using the cell line H2009, which is refractory to miRNA-induced arrest. In agreement with reports from the literature [7, 16], down-regulation of *c-Met* mRNA was specific for *miR-34a*, whereas down-regulation of *CCND3* and *CCNE1* mRNAs were specific for *miR-15a/16* (Fig. 6B and suppl. Fig. S2). In contrast, *CCNE2* mRNA, which contains a target site for *miR-34a* [16], and *CCNA1* mRNA, which contains no target site, were virtually unaffected.

To address the possibility that the synergistic effect was due to an increased number of targets, *CCNE1*, a target unique to *miR-15a/16* (Fig. 6B and suppl. Fig. S2), was knocked down in A549 cells by RNA interference resulting in 80.3±6.6% less *CCNE1* mRNA. We would expect that the synergistic effect would be reduced under these conditions, given that one of the unique targets, *CCNE1*, has been removed. This was indeed the case. The percentage of cells in the G_1-G_0 phase increased to 36% on co-transfection with CCNE1 siRNA and pre-miR-control (Fig. 7A, stippled columns). However, *CCNE1* siRNA had only a low impact on cell cycle arrest of cells overexpressing *miR-15a/16* (Fig. 7A, white columns), which can be explained by the fact that *CCNE1* was already down-regulated by *miR-15a/16*. In contrast, the percentage of A549 cells in G_1-G_0 increased almost two-fold on co-transfection with *pre-miR-34a* and *CCNE1* siRNA relative to cells co-transfected with pre-miR-34a and si control (Fig. 7A, grey columns), suggesting that CCNE1 siRNA and pre-miR-34a act together to induce cell cycle arrest in a more efficient manner. Notably, CCNE1 siRNA in combination with pre-miR-34a (59.0±3.1%; grey column) and CCNE1 siRNA in combination with both pre-miRNAs (61.6±5.3%; black column; p=0.5) induced a G_1-G_0 arrest with the same efficiency. Thus, the synergistic effect exerted by the combined action of miR-15a/16 and miR-34a was clearly abrogated. In contrast, si control in combination with both pre-miRNAs together gave rise to almost two times more cells in G_1-G_0 than si control in combination with pre-miR-34a or pre-miR-15a/16, respectively ($p<0.004$). The observed effects were not cell-line-specific, since comparable results were obtained for A549 (Fig. 7A) and H1299 cells (Fig. 7B). In conclusion, the synergistic effect of miR-15a/16 and miR-34a is due to the fact that more miRNA targets are down-regulated by the combined action of both miRNAs.

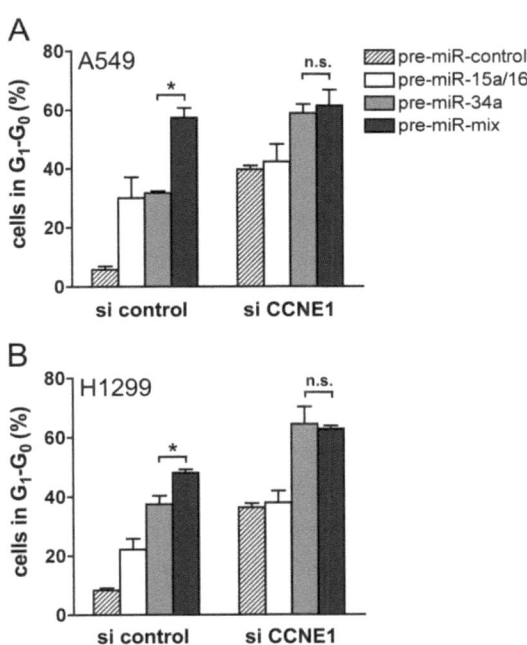

Figure 7. Synergistic action on cell cycle arrest is due to the down-regulation of unique mRNA targets. A549 (A) or H1299 cells (B) were co-transfected with 20 nM miRNA precursors and 7.8 nM siRNA against CCNE1 and subsequently treated for 18 h with nocodazole beginning 24 h post-transfection. Comparable results were also obtained 48 h post-transfection (data not shown).

Discussion

Cell cycle progression critically depends on numerous regulatory processes which are often deregulated in cancer (reviewed by Evan et al. [28]). miRNAs contribute to the complexity of cell cycle control by interfering with a variety of different components of the cell cycle machinery allowing the coordinated regulation of gene expression at the post-transcriptional level (reviewed by Bueno et al. [29] and Carleton et al. [30]).

miR-15a/16 and *miR-34a* share overlapping functions. They both induce cell cycle arrest in G_1-G_0 and share common targets including *CCND1*, *CDK4* and *CDK6*. In addition, the ability of either one of these miRNAs to induce cell cycle arrest in G_1-G_0 largely depends on the expression of Rb (Fig. 3 and ref.[7]). Cyclin D in complexes with CDK4 or CDK6, and cyclin E in a complex with CDK2 regulate progression through the G_1-S boundary of the cell cycle. These complexes phosphorylate and thereby prevent Rb from binding to E2F, which on release, drives cells from G_1 to S phase (reviewed by Morgan et al.[31]). From these results we may conclude that functionally relevant targets of either type of miRNA must be upstream of Rb. These include CCND1, CDK4 and CDK6, which are common to both miRNAs, CCNE1 and CCND3, which are unique to miR-15a/16 and c-Myc and c-Met, which are unique to *miR-34a*. With the exception of CDK4 and c-Myc, all these genes are confirmed targets of *miR-15a/16* and *miR-34a* in NSCLC cells [7, 14, 16, 26]. In contrast, experimentally validated targets downstream of Rb including E2F1, E2F2, E2F3, E2F7, WEE1, CHK1 and CARD10 [25, 32, 33] seem to be less relevant for the regulation of cell cycle progression by *miR-15a/16* or *miR-34a*, at least in NSCLC cells.

The finding that both miRNAs share highly related functions is further illustrated by the fact that both miRNAs are co-regulated in all adenocarcinoma samples. In the majority of NSCLC cases, both miRNAs are significantly down-regulated indicating that they play an important role as tumour suppressor. Tumours can escape the concerted action of *miR-15a/16* and *miR-34a* by down-regulation of both miRNAs or, alternatively, by down-regulation of Rb. Mechanisms which may lead to dysregulation of *miR-15a/16* or *miR-34a* in cancer include deletion of the respective miRNA loci [7, 34], defects in miRNA processing [21], altered promoter methylation [35], or altered expression of transcriptional regulators [36-38]. Defects in miRNA processing may account for only a subgroup of NSCLC, since the majority of tumours either expressed normal or high levels of miR-21. p53 is a potent transactivator of *miR-34a* [39, 40], and is implicated in the processing of *miR-16* [40]. However, no correlation was observed between the mutation status of p53 and the expression level of miR-34a [6] or *miR-15a/16* [41]. In addition, the possibility that both miRNAs are able to mutually regulate their expression can be excluded (Fig. 1C). Thus, it rather seems that several independent mechanisms may account for the dysregulation of *miR-15a/16* and *miR-34a* in NSCLC.

Why is there a relatively high degree of redundancy between miRNAs? To address this question we co-transfected cells with *miR-15a/16* and *miR-34a* and demonstrated that both miRNAs act synergistically to induce cell cycle arrest in G_1-G_0. In contrast, the concerted action of these miRNAs on common mRNA targets was additive rather than synergistic. Thus, there seems to be little interference in binding of these miRNAs to the same target molecule and each miRNA contributes to the mRNA stability in an independent manner. The synergistic effect can rather be explained by the fact that in addition to their targets common to both miRNAs they are also able to bind to targets unique to either type of miRNA. Thus, in a combinatorial mode, both miRNAs are able to down-regulate more targets than each miRNA alone. This is based on the finding that knocking down *CCNE1*, a target unique to *miR-15a/16*, by RNA interference, abrogated the synergistic effect exerted by the combination of both miRNAs (Fig. 7). These effects were not cell-line-specific, since comparable results were obtained with A549 and H1299 cells. This model is in agreement with our results that *miR-34a* and *miR-15a/16* acted synergistically under both saturating and non-saturating conditions. In contrast, if the synergistic effect of these miRNAs were due to a more efficient repression of individual targets, we would expect such an effect to occur only under non-saturating conditions. miRNAs exert fine-tuning regulatory functions, in most cases leading only to a modest repression of target mRNAs and proteins [24]. Our results suggest that miRNAs can potentiate their impact on the regulation of cellular processes by acting in a combinatorial mode.

Surprisingly, we were unable to detect any synergistic effect on apoptosis. Although both miRNAs are able to target *Bcl2*, only *miR-34a* was able to induce apoptosis. This may be due to quantitative differences in their ability to down-regulate *Bcl2*. Alternatively, other anti-apoptotic genes besides *Bcl2*, which are targeted by *miR-34a*, but not *miR-15a/16*, may have to be down-regulated in order apoptosis can occur. It is noteworthy, however, that the observed effects may depend on the cell system as *miR-15a/16* was able to induce apoptosis in CLL [15].

There are only few examples of microRNAs in the literature that act in a synergistic manner. Ivanosvska and Cleary were the first to investigate the concerted action of miR-16 and miR-34a on cell cycle arrest. However, based on their results it was not clear if both miRNAs acted in an additive or synergistic manner [42]. *miR-84* and *let-7* promote terminal differentiation of the hypodermis and cessation of molting in *C. elegans* in a synergistic manner [43]. However, *miR-84* and *let-7* share identical seed sequences, suggesting that they regulate the same set of target genes. In addition, pairs of a cytomegalovirus derived miRNA and a host cell derived miRNA acted on the same gene (*MICB*) through site proximity in a synergistic manner [44]. Thus different mechanisms may exist that may lead to a synergistic action of miRNAs.

Therapeutic strategies for the treatment of human cancer based on modulation of miRNA activity in cancer tissues have gained much attention in the past few years [12, 45-49]. In a recent publication, a new formulation is described that allows the reintroduction of miRNAs, depleted in cancer cells, in order to reactivate cellular pathways that drive a therapeutic response [50]. The authors demonstrated that formulated *miR-34a* blocked tumour growth in a mouse model of NSCLC. Our results suggest that administering formulated *miR-34a* in combination with formulated *miR-15a/16* may lead to a significant increase in the therapeutic impact. This strategy may be particularly effective for the treatment of NSCLC, since both types of miRNAs are normally down-regulated in this class of tumours.

Conclusion

It is generally agreed that miRNAs form part of networks to control cellular processes. Currently, the miRNA field is focused primarily on the identification of novel targets of individual miRNAs, but little information is available how miRNAs act in a combinatorial mode. We show that *miR-34a* and *miR-15a/16* act together to control cell cycle progression in a synergistic and Rb-dependent manner. From these results we may conclude that the combination of miRNAs, which form part of the same network, rather than individual miRNAs should be considered for assessing a biological response. In addition, our study may have translational implications. Since both miRNAs are significantly downregulated in the majority of adenocarcinomas, administering a combination of both miRNAs may potentiate the therapeutic impact of each individual miRNA.

Competing interests

The authors declare that they have no competing interests.

Authors' contribution

NB performed all experiments, participated in the conception and design of the study and helped to draft the manuscript. EV was responsible for the conception of the study and wrote the manuscript. Both authors read and approved the final manuscript.

Acknowledgments

We thank Thomas Brunner for helpful discussions and critical reading of the manuscript. This work was supported by a grant from the Swiss National Science Foundation to EV.

References

1. Edwards BK, Brown ML, Wingo PA, Howe HL, Ward E, Ries LA, Schrag D, Jamison PM, Jemal A, Wu XC, et al: **Annual report to the nation on the status of cancer, 1975-2002, featuring population-based trends in cancer treatment.** J Natl Cancer Inst 2005, **97**:1407-1427.

2. Wistuba, II, Gazdar AF: **Lung cancer preneoplasia.** Annu Rev Pathol 2006, **1**:331-348.

3. Fong KM, Sekido Y, Gazdar AF, Minna JD: **Lung cancer. 9: Molecular biology of lung cancer: clinical implications.** Thorax 2003, **58**:892-900.

4. Hwang HW, Mendell JT: **MicroRNAs in cell proliferation, cell death, and tumorigenesis.** Br J Cancer 2006, **94**:776-780.

5. Takamizawa J, Konishi H, Yanagisawa K, Tomida S, Osada H, Endoh H, Harano T, Yatabe Y, Nagino M, Nimura Y, et al: **Reduced expression of the let-7 microRNAs in human lung cancers in association with shortened postoperative survival.** Cancer Res 2004, **64**:3753-3756.

6. Gallardo E, Navarro A, Vinolas N, Marrades RM, Diaz T, Gel B, Quera A, Bandres E, Garcia-Foncillas J, Ramirez J, Monzo M: **miR-34a as a prognostic marker of relapse in surgically resected non-small-cell lung cancer.** Carcinogenesis 2009, **30**:1903-1909.

7. Bandi N, Zbinden S, Gugger M, Arnold M, Kocher V, Hasan L, Kappeler A, Brunner T, Vassella E: **miR-15a and miR-16 are implicated in cell cycle regulation in a Rb-dependent manner and are frequently deleted or down-regulated in non-small cell lung cancer.** Cancer Res 2009, **69**:5553-5559.

8. Garofalo M, Di Leva G, Romano G, Nuovo G, Suh SS, Ngankeu A, Taccioli C, Pichiorri F, Alder H, Secchiero P, et al: **miR-221&222 regulate TRAIL resistance and enhance tumorigenicity through PTEN and TIMP3 downregulation.** Cancer Cell 2009, **16**:498-509.

9. Frenquelli M, Muzio M, Scielzo C, Fazi C, Scarfo L, Rossi C, Ferrari G, Ghia P, Caligaris-Cappio F: **MicroRNA and proliferation control in chronic lymphocytic leukemia: functional relationship between miR-221/222 cluster and p27.** Blood 2010, **115**:3949-3959.

10. Hayashita Y, Osada H, Tatematsu Y, Yamada H, Yanagisawa K, Tomida S, Yatabe Y, Kawahara K, Sekido Y, Takahashi T: **A polycistronic microRNA cluster, miR-17-92, is overexpressed in human lung cancers and enhances cell proliferation.** Cancer Res 2005, **65**:9628-9632.

11. Takahashi Y, Forrest AR, Maeno E, Hashimoto T, Daub CO, Yasuda J: **MiR-107 and MiR-185 can induce cell cycle arrest in human non small cell lung cancer cell lines.** PLoS One 2009, **4**:e6677.

12. Negrini M, Ferracin M, Sabbioni S, Croce CM: **MicroRNAs in human cancer: from research to therapy.** J Cell Sci 2007, **120**:1833-1840.

13. Hermeking H: **The miR-34 family in cancer and apoptosis.** Cell Death Differ 2010, **17**:193-199.

14. Sun F, Fu H, Liu Q, Tie Y, Zhu J, Xing R, Sun Z, Zheng X: **Downregulation of CCND1 and CDK6 by miR-34a induces cell cycle arrest.** FEBS Lett 2008, **582**:1564-1568.

15. Cimmino A, Calin GA, Fabbri M, Iorio MV, Ferracin M, Shimizu M, Wojcik SE, Aqeilan RI, Zupo S, Dono M, et al: **miR-15 and miR-16 induce apoptosis by targeting BCL2.** Proc Natl Acad Sci U S A 2005, **102**:13944-13949.

16. He L, He X, Lim LP, de Stanchina E, Xuan Z, Liang Y, Xue W, Zender L, Magnus J, Ridzon D, et al: **A microRNA component of the p53 tumour suppressor network.** Nature 2007, **447**:1130-1134.

17. Cannell IG, Kong YW, Johnston SJ, Chen ML, Collins HM, Dobbyn HC, Elia A, Kress TR, Dickens M, Clemens MJ, et al: **p38 MAPK/MK2-mediated induction of miR-34c following DNA damage prevents Myc-dependent DNA replication.** Proc Natl Acad Sci U S A 2010, **107**:5375-5380.

18. Wei JS, Song YK, Durinck S, Chen QR, Cheuk AT, Tsang P, Zhang Q, Thiele CJ, Slack A, Shohet J, Khan J: **The MYCN oncogene is a direct target of miR-34a.** Oncogene 2008, **27**:5204-5213.

19. Tan X, Martin SJ, Green DR, Wang JY: **Degradation of retinoblastoma protein in tumor necrosis factor- and CD95-induced cell death.** J Biol Chem 1997, **272**:9613-9616.

20. Karamitopoulou E, Cioccari L, Jakob S, Vallan C, Schaffner T, Zimmermann A, Brunner T: **Active caspase 3 and DNA fragmentation as markers for apoptotic cell death in primary and metastatic liver tumours.** Pathology 2007, **39**:558-564.

21. Karube Y, Tanaka H, Osada H, Tomida S, Tatematsu Y, Yanagisawa K, Yatabe Y, Takamizawa J, Miyoshi S, Mitsudomi T, Takahashi T: **Reduced expression of Dicer associated with poor prognosis in lung cancer patients.** Cancer Sci 2005, **96**:111-115.

22. Gao W, Yu Y, Cao H, Shen H, Li X, Pan S, Shu Y: **Deregulated expression of miR-21, miR-143 and miR-181a in non small cell lung cancer is related to clinicopathologic characteristics or patient prognosis.** Biomed Pharmacother 2010, **64**:399-408.

23. Zhang JG, Wang JJ, Zhao F, Liu Q, Jiang K, Yang GH: **MicroRNA-21 (miR-21) represses tumor suppressor PTEN and promotes growth and invasion in non-small cell lung cancer (NSCLC).** Clin Chim Acta 2010, **411**:846-852.

24. Baek D, Villen J, Shin C, Camargo FD, Gygi SP, Bartel DP: **The impact of microRNAs on protein output.** Nature 2008, **455**:64-71.

25. Linsley PS, Schelter J, Burchard J, Kibukawa M, Martin MM, Bartz SR, Johnson JM, Cummins JM, Raymond CK, Dai H, et al: **Transcripts targeted by the microRNA-16 family cooperatively regulate cell cycle progression.** Mol Cell Biol 2007, **27**:2240-2252.

26. Liu Q, Fu H, Sun F, Zhang H, Tie Y, Zhu J, Xing R, Sun Z, Zheng X: **miR-16 family induces cell cycle arrest by regulating multiple cell cycle genes.** Nucleic Acids Res 2008, **36**:5391-5404.

27. Reed MF, Zagorski WA, Knudsen ES: **RB activity alters checkpoint response and chemosensitivity in lung cancer lines.** J Surg Res 2007, **142**:364-372.

28. Evan GI, Vousden KH: **Proliferation, cell cycle and apoptosis in cancer.** Nature 2001, **411**:342-348.

29. Bueno MJ, Perez de Castro I, Malumbres M: **Control of cell proliferation pathways by microRNAs.** Cell Cycle 2008, **7**:3143-3148.

30. Carleton M, Cleary MA, Linsley PS: **MicroRNAs and cell cycle regulation.** Cell Cycle 2007, **6**:2127-2132.

31. Morgan DO: **Cyclin-dependent kinases: engines, clocks, and microprocessors.** Annu Rev Cell Dev Biol 1997, **13**:261-291.

32. Tazawa H, Tsuchiya N, Izumiya M, Nakagama H: **Tumor-suppressive miR-34a induces senescence-like growth arrest through modulation of the E2F pathway in human colon cancer cells.** Proc Natl Acad Sci U S A 2007, **104**:15472-15477.

33. Welch C, Chen Y, Stallings RL: **MicroRNA-34a functions as a potential tumor suppressor by inducing apoptosis in neuroblastoma cells.** Oncogene 2007, **26**:5017-5022.

34. Girard L, Zochbauer-Muller S, Virmani AK, Gazdar AF, Minna JD: **Genome-wide allelotyping of lung cancer identifies new regions of allelic loss, differences between small cell lung cancer and non-small cell lung cancer, and loci clustering.** Cancer Res 2000, **60**:4894-4906.

35. Lodygin D, Tarasov V, Epanchintsev A, Berking C, Knyazeva T, Korner H, Knyazev P, Diebold J, Hermeking H: **Inactivation of miR-34a by aberrant CpG methylation in multiple types of cancer.** Cell Cycle 2008, **7**:2591-2600.

36. Hermeking H: **MiR-34a and p53.** Cell Cycle 2009, **8**:1308.

37. Lerner M, Harada M, Loven J, Castro J, Davis Z, Oscier D, Henriksson M, Sangfelt O, Grander D, Corcoran MM: **DLEU2, frequently deleted in malignancy, functions as a critical host gene of the cell cycle inhibitory microRNAs miR-15a and miR-16-1.** Exp Cell Res 2009, **315**:2941-2952.

38. Chang TC, Yu D, Lee YS, Wentzel EA, Arking DE, West KM, Dang CV, Thomas-Tikhonenko A, Mendell JT: **Widespread microRNA repression by Myc contributes to tumorigenesis.** Nat Genet 2008, **40**:43-50.

39. Hermeking H: **p53 enters the microRNA world.** Cancer Cell 2007, **12**:414-418.

40. Suzuki HI, Yamagata K, Sugimoto K, Iwamoto T, Kato S, Miyazono K: **Modulation of microRNA processing by p53.** Nature 2009, **460**:529-533.

41. Navarro A, Diaz T, Gallardo E, Vinolas N, Marrades RM, Gel B, Campayo M, Quera A, Bandres E, Garcia-Foncillas J, et al: **Prognostic implications of miR-16 expression levels in resected non-small-cell lung cancer.** J Surg Oncol 2011, **103**:411-415.

42. Ivanovska I, Cleary MA: **Combinatorial microRNAs: working together to make a difference.** Cell Cycle 2008, **7**:3137-3142.

43. Hayes GD, Frand AR, Ruvkun G: **The mir-84 and let-7 paralogous microRNA genes of Caenorhabditis elegans direct the cessation of molting via the conserved nuclear hormone receptors NHR-23 and NHR-25.** Development 2006, **133**:4631-4641.

44. Nachmani D, Lankry D, Wolf DG, Mandelboim O: **The human cytomegalovirus microRNA miR-UL112 acts synergistically with a cellular microRNA to escape immune elimination.** Nat Immunol 2010, **11**:806-813.

45. Heneghan HM, Miller N, Kerin MJ: **MiRNAs as biomarkers and therapeutic targets in cancer.** Curr Opin Pharmacol 2010, **15**:673-682.

46. Zhang S, Chen L, Jung EJ, Calin GA: **Targeting microRNAs with small molecules: from dream to reality.** Clin Pharmacol Ther 2010, **87**:754-758.

47. Sarkar FH, Li Y, Wang Z, Kong D, Ali S: **Implication of microRNAs in drug resistance for designing novel cancer therapy.** Drug Resist Updat 2010, **13**:57-66.

48. Seto AG: **The road toward microRNA therapeutics.** Int J Biochem Cell Biol 2010, **42**:1298-1305.

49. Petri A, Lindow M, Kauppinen S: **MicroRNA silencing in primates: towards development of novel therapeutics.** Cancer Res 2009, **69**:393-395.

50. Wiggins JF, Ruffino L, Kelnar K, Omotola M, Patrawala L, Brown D, Bader AG: **Development of a lung cancer therapeutic based on the tumor suppressor microRNA-34.** Cancer Res 2010, **70**:5923-5930.

Supplementary figures

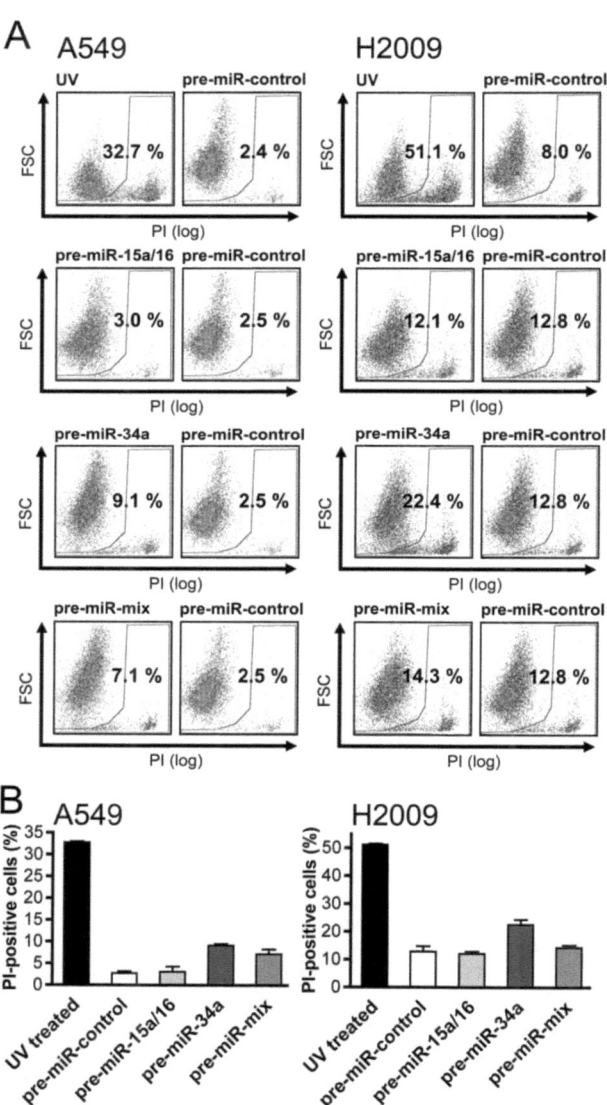

Supplementary figure S1. Analysis of propidium iodide (PI)-stained cells by flow cytometry. H2009 and A549 cells were transfected as described in the legends to Fig. 5 and analysed 72 h or 96 h post-transfection, respectively. (A), dot plot of FSC vs. PI (log) of the transfection experiments in Fig. 5B. (B), percent PI-positive cells. The mean ± SD from independent transfections is presented (n≥3).

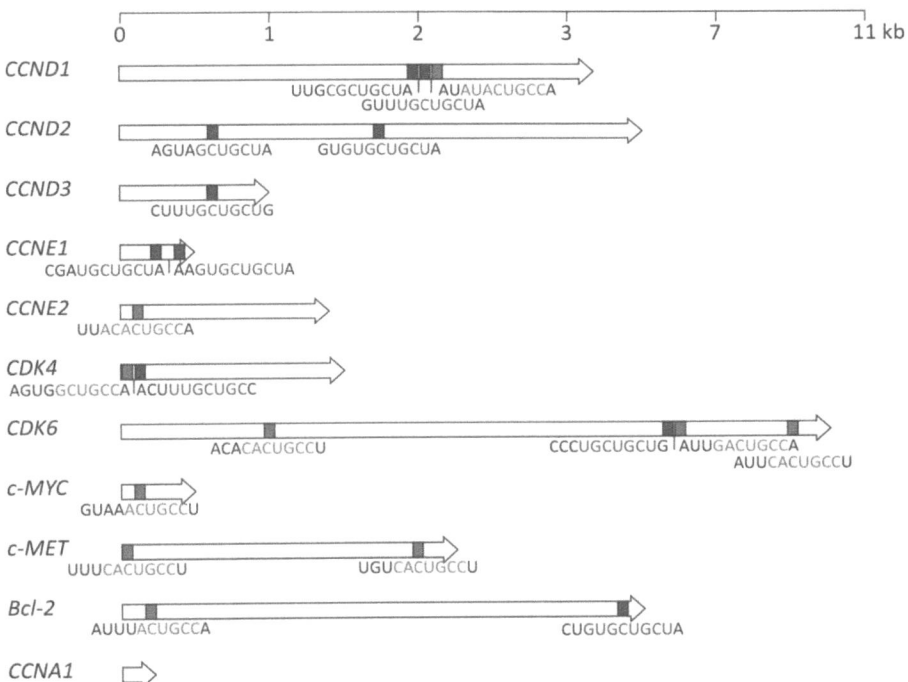

Supplementary figure S2. Schematic depiction of the 3' untranslated region of experimentally validated (*CCND1, CCND2, CCND3, CCNE1, CDK4, c-MET* and *Bcl2*) and predicted (*CDK6* and *c-MYC*) targets of *miR-15a/16* and *miR-34a* in NSCLC cell lines. *miR-15a/16*-specific target sites are highlighted in dark grey and *miR-34a*-specific target sites are highlighted in light grey. CCNA1 contains no *miR-15a/16* or *miR-34a*-specific target sites.

Manuscript 3

Linking microRNAs to aberrant cell cycle control in cancer

Erik Vassella* and Nora Bandi

submitted

will be published by Humana Press 2012
as a chapter in
Methods in Molecular Biology
Cell Cycle Control: Mechanisms and Protocols (2nd Edition)

Institute of Pathology, University of Bern, Bern, Switzerland

*Corresponding author: E. Vassella
Institute of Pathology
University of Bern
Murtenstrasse 31
CH-3010 Bern
Switzerland

phone: 0041-31-632 9943
fax: 0041-31-381 8764
Email: erik.vassella@pathology.unibe.ch

Running head: cell cycle control by microRNAs

Key words: cell cycle control, cancer, formalin-fixed paraffin embedded tissue, target prediction, laser capture microdissection, luciferase activity assay, microRNA, real-time PCR

Linking microRNAs to aberrant cell cycle control in cancer

Erik Vassella and Nora Bandi

Institute of Pathology, University of Bern, Bern, Switzerland

MicroRNAs (miRNAs) are small regulatory RNAs that play key roles in important biological processes including proliferation, differentiation and apoptosis. Accumulating evidence also indicates the involvement of miRNA alterations in human cancer. This chapter describes a protocol for quantification of miRNAs in tumour tissues and matched normal tissues collected by laser capture microdissection. In addition, it describes how a functional analysis of miRNAs can be performed in order to assess if these molecules are directly involved in cell cycle control, allowing to link dysregulation of miRNAs to aberrant cell cycle control in human cancer.

1. Introduction

Tumourigenesis is driven by genetic and epigenetic alterations of oncogenes and tumour suppressor genes leading to the loss of cell cycle control and increased proliferation capacity. microRNAs (miRNAs), which constitute a new class of small non-coding RNAs, have emerged relatively recently as an additional layer of gene expression at the posttranscriptional level. These molecules of 19-23 nucleotides are phylogenetically highly conserved and play key roles in a wide variety of biological processes including cell fate specification, proliferation and apoptosis. Recent evidence indicates that miRNAs also function as tumour suppressors and oncogenes and thus may play an important role in tumour development (1).

miRNAs are first transcribed by RNA polymerase II to primary miRNAs (pri-miRNAs), which are subsequently processed in the nucleus into ~70 nucleotide precursor miRNAs (pre-miRNAs) by the Drosha-DGCR8 complex. The product is subjected to a further cleavage step in the cytoplasm, catalysed by Dicer, giving rise to a double-stranded RNA of ~22 nucleotides. One strand is incorporated into the RNA-induced silencing complex (RISC), where it binds to conserved regions in the 3' untranslated region (3' UTR) of mRNA targets forming imperfect sequence matches. miRNAs can repress translation or degrade their mRNA targets depending on the degree of complementarity to the mRNA sequence (2). However, based on a recent publication, mammalian miRNAs seem to predominantly act on the mRNA level (3).

Despite the fact that miRNAs constitute a relatively small family of less than 1000 genes, they are potentially able to regulate one third of human genes. miRNAs can often bind to hundreds of different targets forming an integrated network which is biased towards certain cellular processes while omitting others (4). It is presumably the summation of different interactions that dictates if a specific pathway is active or not. Involvement of miRNAs in cell cycle control was first revealed by Bernstein et al. in 2003 (5). By establishing a Dicer 1 null mutant, they found a developmental arrest of embryos and loss of stem cells during gastrulation. Since then, miRNAs have emerged as important regulators of different components of the cell cycle machinery (see figure 1).

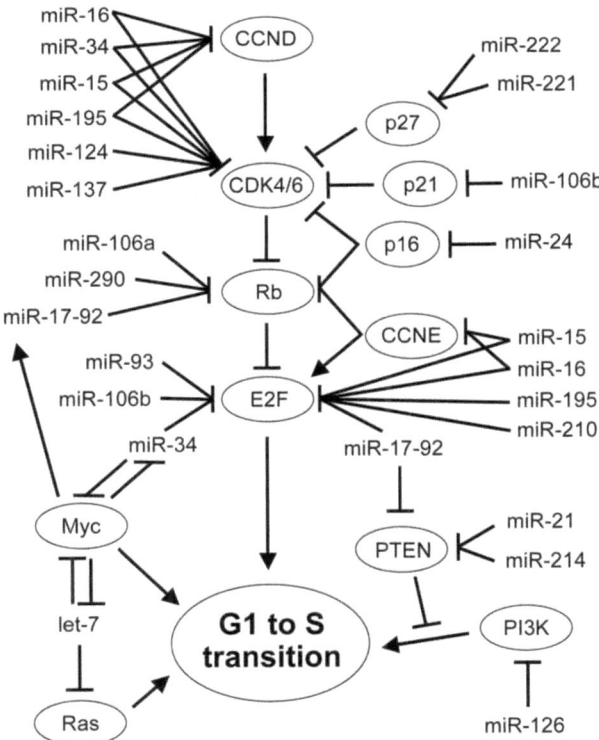

figure 1. integrated network of miRNAs that regulate the G_1 to S transition of the cell cycle

The maybe most prominent example is *let-7*, an important regulator of k-RAS (6). *Let-7* down-regulation is found in some tumours over-expressing RAS protein and correlates with shortened post-operative survival of lung cancer patients (7). Other components including those of the PI3K/AKT and EGFR signalling pathways as well as components of cyclin/CDK complexes and downstream effectors involved in G_1/S or G_2/M checkpoints joined the list of confirmed targets of miRNAs (see figure 1) (8). Those miRNAs with oncogenic properties may exert their function through the inhibition of cell cycle inhibitors (e.g. p27, p21, p16) while those with tumour suppressing properties act on positive regulators of the cell cycle. Interestingly, some miRNAs form part of interaction networks with regulators of the cell cycle resulting in regulatory loops that are frequently altered in human cancer. For example, E2F induces the expression of *miR-17-92* which, in turn, down-regulates E2F (9). An inverse situation has been ascribed for *let-7* and *miR-34*: both miRNAs are able to down-regulate Myc, while the latter protein blocks transcription of either one of these miRNAs (10). Both regulatory loops are linked since Myc is able to induce expression of *miR-17-92* and thereby modulates expression of E2F (figure 1).

MicroRNAs involved in cell cycle control are frequently dysregulated in human cancer: oncogenic miRNAs are frequently over-expressed while tumour suppressing miRNAs are frequently down-regulated. Copy number alterations, mutations interfering with miRNA processing, miRNA stability or binding of miRNAs to target mRNAs, or alterations in the transcription initiation of pri-miRNAs have been attributed as being potential causes for dysregulation of miRNAs in tumour tissues (11). The fact that miRNAs are frequently dysregulated in cancer opens a new field of therapeutic strategies based on modulation of miRNA activity for the treatment of human cancer. In particular, miRNAs involved in cell cycle control that are dysregulated in human cancer may be interesting targets for therapeutic intervention (12).

This chapter illustrates a possible workflow in order to link dysregulation of miRNAs to aberrant cell cycle control in human cancer. The first part describes a protocol for quantification of miRNAs in tumour tissues and matched normal tissues collected by laser capture microdissection. The second part describes a functional analysis of miRNAs in order to assess if they are directly involved in cell cycle control.

2. Materials

1. Chemicals to be used in these procedures should be of the best grade available commercially. Solutions should be prepared with RNase- and nucleic acid-free sterile water, unless indicted otherwise.

2. Nucleic acid-free water can be obtained from different suppliers (e.g. Ambion, Dharmacon). RNases are inactivated by treating nucleic acid-free water with 0.1% diethyl pyrocarbonate (e.g. from Sigma) for 30 minutes followed by autoclavation.
3. miRNA mimics and miRNA inhibitors can be obtained from Ambion, Exiqon, Dharmacon or other companies. Prepare 100 µM stock solutions in RNase-free water and distribute into small aliquots. Store at -80°C.
4. Transfection of cells with miRNA mimics or miRNA inhibitors can be performed using HiPerFect or Attractene transfection reagents (Qiagen), or transfection regents from other companies.
5. Appropriate cell culture medium; e.g. Iscove's Modified Dulbecco's Medium supplemented with 1% penicillin/streptomycin solution, 2 mM L-alanyl-L-glutamin, 10% (v/v) FBS from Sigma-Aldrich, and trypsin (e.g. from Sigma-Aldrich).
6. Nocodazole (e.g. from Sigma-Aldrich): Prepare as a 1 mg/ml stock solution in DMSO and store at -20°C in aliquots.
7. PBS without calcium or magnesium (e.g. from Sigma-Aldrich).
8. Bovine serum albumin (BSA), lyophilized powder (e.g. from Sigma-Aldrich).
9. Propidium iodide (PI, e.g. from Sigma-Aldrich): Prepare as a 1 mg/ml stock solution in water and store at 4°C.
10. RNase A (e.g. from Sigma-Aldrich): Prepare as a 10 mg/ml stock solution in water and store in aliquots at -20°C.
11. PI/RNase A solution: 1 mg/ml RNase A, 200 µg/ml propidium iodide in PBS. Prepare at the same day and keep at 4°C protected from light until use.
12. RNA isolation kit suitable for the isolation of small RNA molecules (e.g. RecoverAll total nucleic acid isolation kit or MirVana miRNA isolation kit from Ambion).
13. Reverse transcription reagents for reverse transcription of small RNAs can be obtained from Applied Biosystems or Qiagen.
14. qPCR primers and reagents to quantify miRNAs and appropriate normalisation controls can be obtained from Applied Biosystems or Qiagen.
15. Yeast tRNA (e.g. from Sigma-Aldrich). Prepare as a 120 ng/ul stock solution in RNase-free water and store in aliquots at -80°C.

16. Membrane slides for laser capture microdissection (e.g PEN membrane slides from Zeiss).
17. Cresyl Violet (Sigma) staining solution. Dissolve 50 mg solid Cresyl Violet in 5 ml 50% ethanol. Stir for several hours or overnight at room temperature and filter the solution by passing through a 0.45 µm filter unit. Cresyl Violet staining solution can be stored at 4°C for up to 4 weeks.
18. Dual Luciferase-assay kit (Promega).
19. Expression plasmids for Renilla Luciferase and Firefly Luciferase reporter assays (e. g. from Promega).

3. Methods

3.1. MiRNA target-predicting algorithms

The miRBase database (www.mirbase.org) is the primary online repository for miRNA sequences and annotation. It provides a searchable database for all published miRNA sequences and annotations and a pipeline for the prediction of miRNA target genes. Web resources containing lists of computationally predicted targets of miRNAs across many species are listed below.

- TARGETSCAN: www.targetscan.org
- MICRORNA.ORG: www.microrna.org
- PICTAR-VERT: pictar.mdc-berlin.de
- MICROCOSM: www.ebi.ac.uk/enright-srv/microcosm
- DIANA – microT: diana.cslab.ece.ntua.gr/microT

Target prediction is based on different algorithms including complementarity between the seed-region of the miRNA and a sequence stretch in the 3' untranslated region of the mRNA target, thermodynamics, binding site structures and sequence conservation. It is noteworthy that not all predicted targets are indeed real targets and therefore have to be confirmed experimentally (see section 3.3.3.). In addition, the ability of a miRNA to bind to its target also depends on the concentration of either of these molecules. Thus, it has to be shown experimentally if a miRNA is able to regulate the expression of a target at its physiological concentration.

3.2. Quantification of miRNAs from fresh-frozen and formalin-fixed tissues

Archives of formalin-fixed, paraffin embedded tissue specimens in histopathological departments are invaluable research resources for studying the molecular basis of diseases. Quantification of RNAs is challenged by extensive fragmentation and modification of nucleic acids during formalin-fixation. However, different validation studies have demonstrated that miRNAs are relatively well preserved in tissue samples after formalin-fixation and paraffin embedding and can be efficiently extracted and evaluated (13, 14). Methylol groups that are introduced into the RNA by formalin-fixation form intra- and intermolecular bridges. These modifications severely affect reverse transcription and lead to higher Ct values of the real time PCR. To reverse methylol groups, we normally subject nucleic acids to a heat treatment before RNA extraction. In order to assess if the integrity of a specific miRNA is preserved following formalin-fixation, it is recommended to compare the steady-state level of each individual miRNA to be analysed between fresh-frozen and formalin-fixed material derived from the same tissue section by real time PCR or by other methods used for miRNA quantification (see below and Note 1).

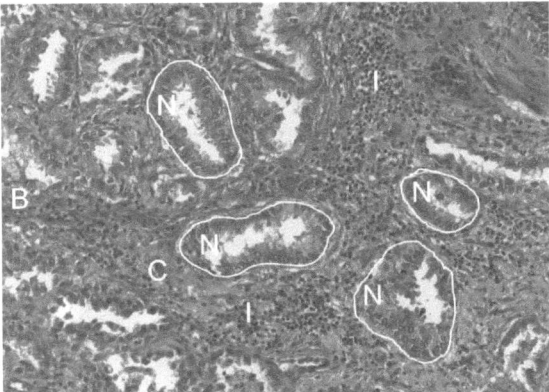

figure 2. Hematoxilin & eosin-stained tissue section of an adenocarcinoma of the lung. Neoplastic cells (N), connective tissue (C), blood vessels (B) and inflammatory cells (I) are indicated.

It is well established that the steady-state level of miRNAs vary from tissue to tissue. Solid tumours comprise two distinct compartments, neoplastic cells and non-malignant stroma including connective tissue, blood vessels and inflammatory cells (see figure 2), which may significantly interfere with miRNA quantification. In some tumours, stroma cells may even outnumber neoplastic cells. Laser capture microdissection (LCM) is an ideal method that enables microdissection of specific cell populations including tumour cells from complex tissues. In addition, matched normal tissues of the tumour samples can also be collected by LCM allowing a direct comparison of the miRNA levels using the ΔΔCt method (Note 2). In a recent study we have compared tumour tissues from adenocarcinomas and squamous cell carcinomas with matched normal tissue from alveolar or bronchial epithelium, respectively, which had been microdissected from the same tissue section, in order to assess if *miR-16* is dysregulated in the tumour samples (15).

Although the short length of mature miRNA is definitely an advantage if quantification is from formalin-fixed tissues, it poses a significant technical challenge for the quantification of such molecules. Thus, conventional RNA purification and detection methods may not be suitable for miRNAs. In addition, discriminating between closely related miRNAs makes the analysis of this class of molecules rather difficult. To date, there are various methods available for discovering, detecting or quantifying miRNAs such as conventional cloning followed by sequencing, deep sequencing, hybridization based methods including microarrays, Northern blot analysis and bead-based hybridization methods, and real-time quantitative assays including Invader assays and RT-PCR assays using either a TaqMan^R probe or SYBR^R green as reporters.

Quantitative real-time PCR is an ideal technique for profiling a limited number of miRNAs if only a small amount of starting material, e.g. laser capture microdissected material, is available for the analysis. This method is linear over a wide concentration range of target miRNAs (Note 3). There exist different reverse transcription kits that are commercially available and methods designed to quantify small RNAs by quantitative real-time PCR. In our laboratory we use TaqMan miRNA assays specific for miRNAs of interest together with the TaqMan microRNA reverse transcription kit and TaqMan universal PCR master mix (Applied Biosystems).

Normalisation of miRNA expression data is necessary to correct sample-to-sample variations. Small nuclear RNAs including *U6, RNU47* and *RNU48* are well suited as normalization controls because (i) they are stably expressed across many tissues, (ii) they are short and therefore behave in a similar way as miRNAs during formalin fixation and RNA extraction and (iii) are expressed at a similar range as miRNAs. In contrast, ribosomal RNAs or mRNAs are much longer, and are often expressed at a much higher level.

Fresh-frozen tissues are of course also suitable for laser capture microdissection and can be used for the analysis of miRNAs. Since a detailed protocol exists for microdissection of fresh-frozen tissues (www.zeiss.com/microdissection), we dedicated ourselves only to a short chapter (3.2.3) on this topic.

3.2.1 Laser capture microdissection of tumour tissue and corresponding normal tissue from FFPE material

1. Carefully clean the microtome workstation and cut 6 µm tissue sections with a fresh knife. Float out tissue section in a warm water bath and mount on a UV-treated PEN 1.0 membrane slide (Note 4).
2. Incubate tissue slides at 58°C for 2 hrs to adhere the tissue to the membrane.
3. Dewax sections by immersing for 2 x 2 min in xylene, then 2 min each in 100% ethanol, 96% ethanol, and 70% ethanol.
4. Wash tissues for 30 sec in RNase-free water and stain with 200 µl of Cresyl Violet staining solution for 30 sec (see Note 5), rinse for 5 sec with RNase-free water, and dehydrate for 30 sec in each 70% ethanol, 95% ethanol, and 100% ethanol.
5. Dry slides at room temperature for 10 min and at 50°C for 10 min.
6. We use the Palm MicroBeam laser system (Zeiss) for microdissection, but other systems can also be used. The focus and energy for cutting and catapulting should be optimized for each type of tissue and the thickness of the section.
7. Place the section for laser microdissection on the microscope stage, mark regions of the tumour within the tissue section and activate the laser (see Note 6). The dissected material is catapulted onto the mineral oil-coated cap of a PCR tube (see Note 7).
8. Resuspend the dissected material in 20 µl of the appropriate digestion buffer for RNA isolation (see below) and collect it by centrifugation at maximal speed for 3 min. Store at - 80°C until use.

3.2.2 RNA isolation from FFPE tissues

1. Contaminations with RNases should be avoided. We recommend cleaning the lab bench with RNase decontamination solution (e.g. RNase-away from Molecular BioProducts.), wearing powder-free laboratory gloves, which should be frequently changed, and using RNase-free plastic-ware and filter pipette tips.

2. We use the RecoverAll total nucleic acid isolation kit from Ambion for isolating miRNAs from FFPE tissues. Adjust the volume of digestion buffer to 133 µl and add 1 µl of a solution containing 120 ng/µl yeast tRNA (see Note 8).
3. Incubate samples for 10 min at 90°C in order to remove methylol groups which may interfere with miRNA quantification.
4. Subject tissue samples to protease digestion and isolate the total RNA fraction by loading on a filter cartridge following the manufacturer's protocol.
5. Elute the RNA by pipetting 30 µl RNase-free water to the centre of the filter following centrifugation for 1 min at maximal speed. Repeat elution step using another 30 µl aliquot of RNase-free water and combine both eluates. Transfer the eluate to a low-binding eppendorf tube and store at -80°C until use.

3.2.3 Laser capture microdissection and isolation of miRNAs from fresh-frozen tissues

1. A detailed protocol exists for laser capture microdissection of fresh-frozen tissues (www.zeiss.com/microdissection). Most pathological departments use Tissue-Tek® O.C.T™ compound as freezing medium, which is suitable for sectioning using a microtome. It is noteworthy, however, that the O.C.T™ should be removed before laser cutting. A similar procedure can be envisaged to microdissect fresh-frozen tissues as described for FFPE tissues.
2. The Qiagen RNeasy micro kit described in the protocol from Zeiss is not suitable for miRNA extraction as this results in a significant loss of small RNA molecules. Instead, the MirVana miRNA isolation kit from Ambion may be more suitable for the isolation of short RNA molecules. We recommend adding yeast tRNA to the RNA extract as this may reduce loss of RNAs owing to unspecific binding. Since RNAs from fresh-frozen tissues contain no methylol groups, the demodification step can be omitted.

3.2.4 Quantification of miRNAs by real-time PCR

All reagents we normally use for quantitative real-time PCR are from Applied Biosystems, but reagents from other companies may also provide satisfactory results.

1. Reverse Transcription (RT) can be carried out in a 15 µl reaction mixture containing 1 mM of each dNTP, 3 µl of RT miRNA primer, 1.5 µl of 10x RT buffer, 3.8 U of RNase inhibitor, 50 U of multiscribe reverse transcriptase and 5 µl of the undiluted RNA

sample. Incubate the reaction mix for 30 min at 42 °C. Inactivate the enzyme by heating the reaction mix for 5 min at 85°C.

2. Dilute the RT product 1: 3 with water and subject 4 µl to a PCR amplification in a 20 µl reaction mixture containing 10 µl of 2x universal mastermix and 1 µl of 20x TaqMan miRNA primers. Start the thermocycling program with an initial 10 min denaturation step at 95°C, followed by 40 cycles (15 sec at 95°C and 60 sec at 60°C). Analyse the results using the ΔΔCt method (see Note 2 and 9).

3.3. Over-expression or silencing of miRNAs in cancer cell lines

To draw a link between dysregulation of certain miRNAs in tumour tissues and aberrant cell cycle control, miRNAs can be over-expressed in tumour cell lines by transfection of double-stranded RNA molecules (miRNA mimics) that mimic Dicer cleavage products. Conversely, 2'-O-methyl-modified oligonucleotides that are complementary to the target miRNA (miRNA inhibitors) can be used to silence endogenous miRNAs. Transfected cells can then be analysed for cell cycle progression using different methods, one of which is described in section 3.3.2. Transfection of miRNA mimics or inhibitors is normally very efficient and may provide results in a very short period of time. However, this method suffers from the drawback that the effect exerted by these molecules disappears after 2-4 days. Therefore these molecules are not suitable for long-term studies in cultured cells. The latter type of experiments depends on DNA plasmids that continuously generate functional miRNAs from endogenous or viral promoters. Introducing miRNA sequences that are flanked by at least 40 nt from their precursor into DNA plasmids is, in general, sufficient to yield mature miRNAs (16). Conversely, plasmids expressing the complementary sequence of a miRNA can be used for silencing of this miRNA. Such plasmids can be introduced into retroviral or adenoviral systems to overcome low transfection efficiency. Many lentiviral vectors over-expressing or silencing a large variety of miRNAs are commercially available.

3.3.1 Transfection of tumour cell lines with miRNA mimics or miRNA inhibitors

In our hands, HiPerfect transfection reagent (Qiagen) is very suitable for transfection of adherent cells with miRNA mimics or miRNA inhibitors, but other transfection reagents exist that may also provide satisfactory results (see Note 10). Depending on the planned experiment for phenotypic analysis, the scale of transfection has to be adjusted. We normally seed cells in 6 cm plates, if cells are analysed by flow cytometry. In contrast, growth experiments can be performed in 96 well plates. Transfection conditions may vary depending on the cell density. Titrate the cell density by trying different confluences of adherent cells. Optimally, cells should reach 70-80% confluence (not higher!) at the day of harvest.

1. Seed cells 1 day prior to transfection. For 24 well plates we use, in general, 1.5×10^4 A549 cells (NSCLC cell line) in 500 µl of medium if cells are harvested 48 hrs post-transfection.

2. At the day of transfection prepare the transfection reaction mix according to the manufacturer's protocol and add the complex to the cells (~20 to 50% confluent). For 24-well plates we use, in general, 0.1-20 nM miRNA mimics or 0.1-100 nM miRNA inhibitors in 137 µl of culture medium. Aspirate the medium over the cells and replace with 166 µl fresh medium. Add 137 µl of the transfection reaction mix dropwise to the cells.

3. After 4-6 hours, add normal growth medium without removing the transfection mixture. For 24-well plates, we add 300 µl growth medium. If toxicity is a problem, remove the transfection mixture and replace with 500 µl of normal growth medium.

4. Analyse the cells using the appropriate protocol 24-72 hours post transfection (see Note 11).

3.3.2 Cell cycle analysis of miRNA precursor-/miRNA inhibitor-transfected cells

The procedure described below is a modification of the method by Linsley et al. (17). This method is routinely used in our laboratory (15) and proved to be useful for the analysis of specific miRNAs for their role in cell cycle control. The results from the expression analyses (see 3.2) may be useful for the design of the transfection experiment: a cell line over-expressing a specific miRNA may not necessarily exhibit a cell phenotype upon transfection with a miRNA mimic since target mRNAs may already be maximally repressed by the endogenous miRNA. Conversely, it may not be recommended to transfect a cell line with a miRNA inhibitor, if the endogenous miRNA is already severely down-regulated.

The G_1/G_0 arrest phenotype exerted by a miRNA mimic or miRNA inhibitor may be more pronounced if cells are treated with nocodazole 24-48 hrs post-transfection. We have shown previously that transfection of A549 cells with *miR-16* mimic results in a

G_1/G_0 arrest. When cells were cultured in the absence of nocodazole and analysed by flow cytometry 24 hrs post transfection, 69% of the population was in this phase of the cell cycle (see figure 3A, upper panel). However, not all cells in G_1/G_0 had undergone cell cycle arrest. This is based on the finding that transfection with scrambled control (53%, figure 3A, lower panel) or untransfected cells (data not shown) also give rise to a significant proportion of the population in G_1/G_0 phase although these cells are clearly not arrested. The percentage of cells, which had indeed undergone arrest can be determined by treating them with nocodazole for 20 hrs beginning at 24-48 hrs post-transfection. Under these conditions, 33% of the population transfected with *miR-16* was in G_1/G_0 (figure 3B, upper panel) while only 8% of the population transfected with miRNA scrambled control was in this phase of the cell cycle (figure 3B, lower panel).

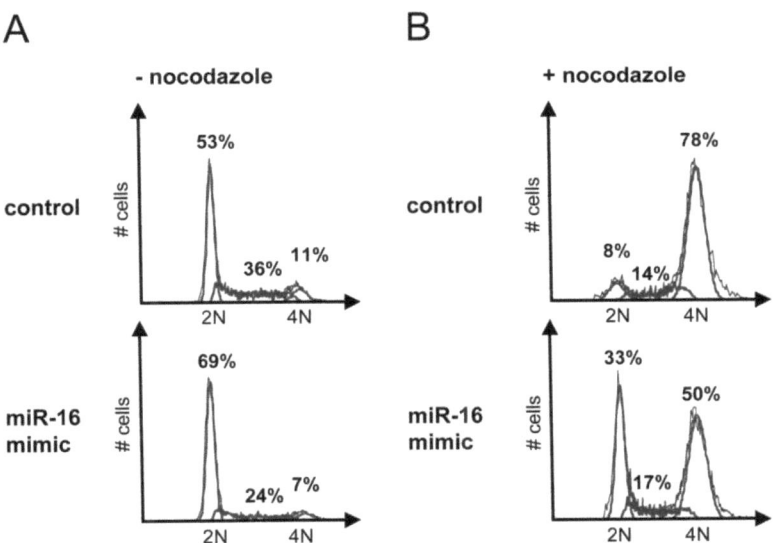

figure. 3. Cell cycle analysis of propidium iodide-stained A549 cells transfected with scrambled control (upper panel) or *miR-16* mimic (lower panel). (A) Cells were cultured in the absence of nocodazole and analysed by flow cytometry 24 hrs post-transfection. (B) Cells were treated with nocodazole for 20 hrs beginning at 24 hrs post-transfection

By analogy, the phenotype of a miRNA, which induces arrest in G_2/M, may be more pronounced if cells are treated with mimosine or aphidicoline, inducers of arrest in late G_1 phase or early S phase, respectively.

Manuscript 3

1. Transfect cells with miRNA mimics and add 100 ng/ml nocodazole 24-48 hrs post-transfection. Continue to culture for 16-20 hrs.
2. Harvest floating and adherent cells and combine them (see Note 12). Collect cells by centrifugation for 5 min at 500 x g.
3. Aspirate the supernatant and wash the cell pellet with 1 ml PBS.
4. Centrifuge again and carefully resuspend the pellet in 300 µl of PBS. It is important to obtain a single-cell suspension prior to fixation.
5. Add 700 µl of ice-cold ethanol absolute (stored at -20°C till use) dropwise to the cell suspension while vortexing to prevent cell clumping. Fix overnight at 4°C (see Note 13).
6. Collect cells by centrifugation at 500 x g for 10 min. Aspirate the supernatant.
7. Vortex cell pellet and wash it with 1 ml PBS containing 1% BSA (see Note 14).
8. Collect cells by centrifugation and carefully resuspend the pellet in 50 µl PI/RNaseA staining solution. Incubate cells at 37°C for 30 minutes.
9. Add 200-500 µl PBS to the stained cells and remove cell clumps by passing the cell suspension through a Cell Strainer FACS filter tube (BD Biosciences).
10. Analyse at least 10'000 cells by flow cytometry.

3.3.2 Identification of miRNA targets

Several independent groups have established computational algorithms designed to predict target genes of miRNA sequences (see 3.1). To date, several methods have been established to show miRNAs regulating their putative target genes. Most commonly used are luciferase reporter constructs containing the target 3' untranslated region (3' UTR) with the putative binding site downstream of the coding region of the reporter gene. Introducing sequences of the target gene that are flanked by at least 40 nt from the miRNA binding site downstream of the luciferase gene is, in general, sufficient to achieve a direct interaction of a miRNA with its target. A fragment encompassing the miRNA binding site can be obtained by PCR using primers flanking the putative binding site and genomic DNA as a template. Alternatively, complementary oligonucleotides encompassing these sequences can be synthesized which, following annealing, can be cloned into an appropriate restriction site downstream of the luciferase gene.

These constructs are used to transfect cells expressing the relevant miRNA. If a construct containing a putative binding site gives rise to a lower luciferase activity as compared to the parental plasmid lacking this site, this indicates that this is a direct target. As a final prove to show that this is indeed the case, mutations should be introduced into the target site of the luciferase construct either by site-directed mutagenesis or, alternatively, by the synthesis of complementary oligonucleotides and analysed for luciferase activity. Constructs carrying such mutants are expected to give rise to a luciferase activity similar to that of the parental plasmid.

A complementary approach to assess, if a miRNA is able to regulate its target at physiological concentrations, is to co-transfect luciferase constructs containing a miRNA binding site with miRNA inhibitors. We would expect that this gives rise to a higher luciferase activity as compared to cells co-transfected with the luciferase construct and a scrambled inhibitor control since this would result in silencing of the endogenous miRNA. These types of experiments are important since they demonstrate if a mRNA of interest is indeed a physiological target of a specific miRNA.

Luciferase assays

Luciferase assays are based on the bioluminescent measurement of luciferase activity. The intensity of light emission is linearly related to the amount of luciferase, which can be measured using a luminometer. We recommend performing co-transfection experiments using a firefly luciferase construct containing the miRNA binding site and a renilla luciferase control plasmid which is used for normalisation of firefly luciferase activity. The activities of firefly and renilla luciferases are measured sequentially from a single sample using the dual-luciferase reporter assay (Promega). The advantage of using this system is that it is very sensitive and gives rise to little variations between independent samples.

We also recommend testing the efficiency of co-transfection experiments using a red fluorescence protein (RFP) reporter construct and siGLOGreen (Dharmacon). The transfection efficiency can be monitored using a fluorescence microscope. We routinely use the Attractene transfection reagent (Qiagen), which is suitable for co-transfection of plasmid DNA in combination with short RNA molecules.

1. Subject cells to co-transfection of plasmid DNA in combination with miRNA mimics or miRNA inhibitors (see Note 15) in a 24 well plate as outlined in section 3.3.1.

Depending on the cell type we usually use 2-15 ng firefly luciferase reporter plasmid, 10-60 ng renilla luciferase reporter plasmid, 0.1-20 nM miRNA mimic and/or 0.1-100 nM of miRNA inhibitor for transfection (see Note 16 and 17).

2. Following transfection, incubate cells at 37°C/5% CO_2 in a humidified incubator for 24-48 h.
3. Aspirate the medium, and gently wash the cells with 0.5 ml PBS.
4. Directly lyse the cells in the well by pipetting 100 µl of lysis buffer to each well (see Note 18 and 19)
5. Transfer 10 µl of the lysate to a Corning 96 well plate (NUNC) and perform the luciferase assay according to the manufacturer's protocol (Promega). We routinely measure luciferase activity in an Infinite 200 Luminometer (Tecan) at 500 ms integration time.

4. Notes

1. The quality of miRNAs may also depend on the post-operative period before fixation. To investigate if this affects the steady-state level of *miR-16*, we kept experimental tissue samples at room temperature for up to 5 hrs before fixation, which, however, did not affect *miR-16* quantification.
2. The comparative $\Delta\Delta Ct$ ($2^{-\Delta\Delta Ct}$) method is used to calculate changes in gene expression as a relative fold difference between an experimental and calibration sample, e.g. tumour tissue vs. normal tissue.
3. *miR-16* detection is linear in a range between 0.05 pg and 500 pg of total RNA.
4. Treatment of membrane-slides with UV-Light (254 nm) for 30 min improves binding of tissues to the membrane and removes contaminating nucleic acids.
5. In our hands, Cresyl Violet has no influence on the quality of RNA in the tissue sample.
6. We normally microdissect approximately 1000 tumour cells per sample. However, the minimal area of cells to be microdissected depends on the expression level of the miRNA of interest.
7. The cap of a PCR tube is coated using 1 µl mineral oil (Sigma).

8. Addition of yeast tRNA to the samples reduces loss of material owing to unspecific binding.
9. Negative controls for real-time PCR should include a no-RT control, and no template controls for reverse transcription and PCR amplification.
10. Only methods that are also suitable for transfection of siRNAs should be used for transfection of miRNA mimics or miRNA inhibitors. Transfection efficiency may vary depending on the cell line and should be optimized for each cell line using fluorescently labelled oligonucleotides (e.g. siGLO Green, Dharmacon). In addition it may be advantages to titrate down the concentration of miRNA mimics or miRNA inhibitors in order to select the minimal concentration inducing a biological effect. Under these conditions, minimal off-target effects would be expected. In addition, artefacts such as saturating available RISC complexes owing to massive over-expression of miRNA mimics can be prevented.
11. It may be recommended to perform a time course in order to determine the optimal time point for the experiment.
12. We recommend using at least 200 000 cells for ethanol fixation.
13. Cells can be stored in the fixative for up to 3 weeks.
14. Ethanol-fixed cells are not efficiently pelleted by centrifugation. This can be overcome by the addition of 1% BSA to the wash medium.
15. Transfection efficiency of plasmid DNA and miRNA mimics/inhibitors may depend on whether cells are transfected with either one of these molecules alone or co-transfected with a combination of both types of molecules. Thus, transfection efficiency may have to be reoptimized in co-transfection experiments.
16. In order to reduce variations of luciferase activity between independent transfections it may be recommended to prepare master mixes for transfection reagents, plasmid DNA and miRNA mimics/inhibitors.
17. High concentrations of firefly reporter plasmid may overwrite the effect exerted by miRNA mimics or miRNA inhibitors, respectively. It is therefore recommended to titrate down the amount of plasmid DNA for transfection.
18. Put the 24-well plate on a shaker and shake vigorously until cells are completely lysed.
19. Luciferase activity in the lysates is stable for up to one year if stored at -80 °C.

Refereces

1. Hwang HW, Mendell JT (2006) MicroRNAs in cell proliferation, cell death, and tumorigenesis. Br J Cancer 94:776-780
2. Zeng Y (2006) Principles of micro-RNA production and maturation. Oncogene 25:6156-6162
3. Guo H, Ingolia NT, Weissman JS et al (2010) Mammalian microRNAs predominantly act to decrease target mRNA levels. Nature 466:835-840
4. Hornstein E, Shomron N (2006) Canalization of development by microRNAs. Nature Genetics 38:20-24
5. Bernstein E, Kim SY, Carmell et al (2003) Dicer is essential for mouse development. Nat Genet 35:215-217
6. Johnson SM, Grosshans H, Shingara J et al (2005) RAS is regulated by the let-7 microRNA family. Cell 120:635-647
7. Takamizawa J, Konishi H, Yanagisawa K et al (2004) Reduced expression of the let-7 microRNAs in human lung cancers in association with shortened postoperative survival. Cancer Res 64:3753-3756
8. Chen D, Farwell MA, Zhang, B (2010) MicroRNA as a new player in the cell cycle. J Cell Physiol 225:296-301
9. Sylvestre Y, De Guire V, Querido E et al (2007) An E2F/miR-20a autoregulatory feedback loop. J Biol Chem 282:2135-2143
10. Sampson VB, Rong NH, Han J et al (2007) MicroRNA let-7a down-regulates MYC and reverts MYC-induced growth in Burkitt lymphoma cells. Cancer Res 67:9762-9770
11. Calin GA, Croce CM (2006) MicroRNA signatures in human cancers. Nat Rev Cancer 6:857-866
12. Wang V, Wu W (2009) MicroRNA-based therapeutics for cancer. BioDrugs 23:15-23
13. Nelson PT, Baldwin DA, Scearce LM et al (2004) Microarray-based, high-throughput gene expression profiling of microRNAs. Nat Methods 1:155-161
14. Jay C, Nemunaitis J, Chen P et al (2007) miRNA profiling for diagnosis and prognosis of human cancer. DNA Cell Biol 26:293-300
15. Bandi N, Zbinden S, Gugger M et al (2009) miR-15a and miR-16 are implicated in cell cycle regulation in a Rb-dependent manner and are frequently deleted or down-regulated in non-small cell lung cancer. Cancer Res 69:5553-5559
16. Krutzfeldt J, Poy MN, Stoffel M (2006) Strategies to determine the biological function of microRNAs. Nat Genet 38:14-19
17. Linsley PS, Schelter J, Burchard J et al (2007) Transcripts targeted by the microRNA-16 family cooperatively regulate cell cycle progression. Mol Cell Biol 27:2240-2252

5. Discussion

Lung cancer is characterised by multiple and heterogeneous genetic alterations. Recent evidence indicates that miRNAs play important roles in human carcinogenesis and can act either as oncogenes or tumour suppressors. Changes in the miRNA expression have been shown to deregulate cancer-related genes involved in several processes as diverse as cell cycle control and apoptosis.

The present work identifies *miR-15a/16* and *miR-34a* as negative regulators of cell cycle progression in NSCLC.

We demonstrate that *miR-15a/16* and *miR-34a* are co-regulated in the majority of squamous cell carcinomas and adenocarcinomas of the lung. In most cases, both miRNAs are significantly down-regulated, indicating that they play an important role as tumour suppressor in NSCLC.

Interestingly, both miRNAs are up-regulated in some cancer types: *miR-15a/16* are frequently over-expressed in cervical cancer [157], while *miR-34a* up-regulation has been observed in pedriatic acute lymphoblastic leukaemia [158] and liver cancer [156]. This may suggest that, depending on cellular context, *miR-15a/16* and *miR-34a* can either act as oncogene or tumour suppressor gene. A dual role as oncogenes and tumour suppressors has also been ascribed for protein-coding genes including *p53, NFAT and E2F1* [159-161]. Consequently, a similar mechanism may also exist in NSCLC, because both miRNAs were up-regulated in 18% of adenocarcinomas. Remarkable, both miRNAs are co-regulated in all adenocarcinoma samples while we observed a differential expression in some of the analysed squamous cell carcinomas. In these tumours *miR-15a/16* was up-regulated in 33%, while *miR-34a* was down-regulated in all cases. Interestingly, an inverse regulation of either one of these miRNAs has been found in liver cancer, where *miR-34a* seems to be up- and *miR-15a/16* down-regulated [156]. These findings are in accordance with previous reports which have shown that different (sub-) types of cancer can be distinguished by their different miRNA expression patterns [62]. Thus depending on cellular context *miR-15a/16* and *miR-34a* may be involved in different regulatory networks in the tumourigenesis of different cancer (sub-) types.

The underlying molecular mechanisms leading to co-regulation of these miRNA are not clearly understood. Down-regulation of these miRNAs is not directly linked since neither *miR-15a/16* nor *miR-34a* are able to affect the expression of each other. Co-repression of

Discussion

these miRNAs may be due to defects in miRNA processing. Notably reduced expression of Dicer has been detected in NSCLC [122]. However, the fact that *miR-21* is up-regulated or expressed at normal levels in the same NSCLC tissues excludes the possibility that defects in miRNA processing alone can lead to the observed co-regulation of *miR-15a/16* and *miR-34a*. The transcription factor p53 is a potent inducer of *miR-34a* [162] and to some extend is also implicated in the maturation of *miR-16* [163]. However, no correlation has been observed between mutation status of p53 and *miR-34a* in NSCLC [91]. *p53* mutations occur in 35% of adenocarcinomas and 50% of squamous cell carcinomas [164]. Thus, genetic alterations of p53 may account for miRNA down-regulation in only a subgroup of NSCLC. In addition, reintroduction of this transcription factor into p53 deficient H2009 cells only resulted in increased levels of *miR-34a*, while *miR-15a/16* remained unaffected (data not shown). Thus it is unlikely, that co-regulation of *miR-15a/16* and *miR-34a* is due to p53 mutations. It is conceivable that additional mechanisms, including altered expression of other transcription factors, genomic rearrangements and deletions of both miRNA loci, point mutations or/and altered promotor methylations may contribute to the deregulation of these miRNAs in tumour cells. In summary, it seems that several independent mechanisms may account for the dysregulation of *miR-15a/16* and *miR-34a* in NSCLC.

We and others have shown, that *miR-15a/16* and *miR-34a* induce a cell cycle arrest in G_1/G_0. Thus one way to escape miRNA-induced arrest is to down-regulate both of these miRNAs. But how can tumour cells grow, when the levels of these miRNAs are high? We demonstrate, that the inactivation of Rb is an alternative mechanism by which NSCLC cells can escape the growth inhibitory signals of *miR-15a/16* and *miR-34a*. This was evidenced by our findings, that H2009 cells lacking Rb are completely resistant to miRNA-induced arrest, whereas introduction of a functional copy of Rb renders them more sensitive. Consistent with these results, down-regulation of Rb in A549 cells by RNA interference confers resistance to those miRNAs, what may apply for up to 14% of squamous cell carcinomas and 33% of adenocarcinomas, which are Rb negative [165]. Consequently, a similar mechanism may also exist in cervical cancer cells in which *miR-15a/16* is frequently up-regulated [157]. Because these cells normally express an inactive form of Rb, we may conclude that the cell cycle progression of cervical cancer no longer depends on *miR-15a/16* activity.

miRNAs can affect hundreds of mRNAs, which makes it difficult to identify the biological relevant targets. Many recent publications demonstrated that important cell cycle genes are directly regulated by either *miR-34a*, *miR-15a/16* or by both of them.

CCND in complexes with CDK4 or CDK6, and CCNE in complex with CDK2 regulate progression through the G_1-S boundary of the cell cycle. These complexes phosphorylate and thereby prevent Rb from binding to E2F, which on release, drives cells from G_1 to S phase (reviewed by ref [166]). Based on our finding that the *miR-15a/16* and *miR-34a* induced G_1/G_0 cell cycle arrest depends on Rb, we may conclude that in NSCLC functionally relevant direct targets of either type of these miRNAs must be upstream of Rb. These include *CCND1*, *CDK4* and *CDK6*, which are common for both miRNAs, *CCNE1* and *CCND3*, which are unique to *miR-15a/16* and *c-Met*, which is unique to *miR-34a*. With the exception of *CDK4*, all these targets were shown to be direct targets of these miRNAs in NSCLC [83, 85, 92, 93]. In addition we show that *miR-15a/16* directly regulate important G_1 cyclins like *CCND1*, *CCND2* and *CCNE1* under physiologic conditions in NSCLC [83]. This is an important finding because, for most validated miRNA-targets, it is still unclear if they are also regulated at physiological concentrations of endogenous miRNAs in specific cell types.

Currently, research is focused primarily on identifying novel targets of individual miRNAs, while the analysis of combinatorial effects of deregulated miRNAs in specific cancer types is still not uncovered. Although many miRNAs share overlapping functions and form part of integrated networks (see chapter 2.2.7), little information is available if they are able to act together to control biological processes. *miR-15a/16* and *miR-34a* are functionally related. Despite the fact that they contain distinct seed sequences, they share common targets and are implicated in similar processes including cell cycle control and apoptosis. Thus, using the example of *miR-15a/16* and the closely related *miR-34a*, we investigated how miRNAs with overlapping functions affect biological processes in a combinatorial mode in NSCLC cells.

Our results indicate that *miR-15a/16* and *miR-34a* act synergistically in inducing cell cycle arrest. Interestingly, the concerted action of these miRNAs on common targets is additive rather than synergistic. Thus there seems to be little interference in binding of these miRNAs to the same target molecule and each miRNA contributes to the mRNA stability in an independent manner. However, the synergistic effect can be explained by the fact that in addition to their common targets, both miRNAs are able to repress targets unique to

Discussion

either type of miRNA. Thus in a combinatorial mode, the two miRNAs are able to down-regulate more targets than each miRNA alone. This is based on the finding that knocking down *CCNE1*, a unique target of *miR-15a/16*, by RNA interference, abrogated the synergistic effect exerted by the combination of those miRNAs.

miRNAs exert fine-tuning regulatory functions, in most cases leading only to a modest repression of target mRNAs [167]. The ability of multiple miRNAs to potentate their impact on the regulation of cellular processes by acting in a combinatorial mode, engenders a novel way of thinking: Instead of a one-miRNA-one–target model, a model with multiple miRNAs that affect multiple targets in a pathway should be considered, since this may result in a potent inhibition of a specific pathway.

Despite the striking similarities of *miR-34a* and *miR-15a/16* in cell cycle control, they differ in their ability to induce apoptosis in NSCLC. Although both miRNAs are able to target *Bcl2*, only *miR-34a* is able to efficiently induce cell death in different NSCLC cell lines. It is noteworthy, however, that the observed effects may depend on the cell system as *miR-15a/16* was able to induce apoptosis in CLL [86]. It would be interesting to determine whether these differences are dictated by the cellular environment or by the cell-type specific expression of other targets involved in apoptosis.

Therapeutic strategies for the treatment of human diseases based on modulation of miRNA activity have gained much attention in the past few years [149-154]. In a recent publication, a new formulation is described that allows the reintroduction of miRNAs, depleted in cancer cells, in order to reactivate cellular pathways that drive a therapeutic response [118]. The authors demonstrated that if administrated intravenously, formulated *miR-34a* blocked tumour growth in a mouse model of NSCLC. Our results suggest that if formulated *miR-34a* is delivered in combination with formulated *miR-15a/16*, this may lead to a significant increase in the therapeutic impact.

In conclusion our results indicate that *miR-15a/16* and *miR-34a* are synergistically implicated in cell cycle control and likely contribute collectively to the tumourigenesis of NSCLC. We propose two alternative pathways by which NSCLC cells escape *miR-15a/16*- and *miR-34a*-induced cell cycle arrest: (i) co-repression of *miR-15a/16* and *miR-34a* or (ii) inactivation of the Rb gene.

Discussion

In addition our results suggest that the combination of miRNAs, which form part of the same network, should be considered for assessing a biological response rather than individual miRNAs. Since *miR-34a* and *miR-15a/16* are frequently down-regulated in the same tumour tissue, these findings may have important therapeutic implications for microRNA-based treatment of Rb-proficient NSCLC.

6. Outlook

It is noteworthy that most miRNAs with a clear tumour suppressor role (e.g. *miR-16*, *miR-29*, and *let-7*) are encoded by more than one gene, which are often located on different chromosomes. Although they are transcribed from distinct precursors, their mature miRNA sequence is identical. Gene duplication may be a way to preserve the function of an important tumour suppressor miRNA if one allele is deleted or silenced. In the present work the steady-state level of other miRNA family members like *miR-15b*, *miR-34*b and *miR-34*c was not addressed. These miRNAs probably contribute to cell cycle control by regulating the same set of targets in NSCLC, although subtle differences between the different family members may exist. In addition, other miRNAs are also involved in similar processes (see figure 5 in the introduction). For example *miR-26a* is related to *miR-15a/16* and *miR-34a* since it also targets cell cycle genes important for G_1 to S transition and induces apoptosis in liver cancer cells [156]. *miR-124* also regulates the transition from G_1 to S phase by the direct down-regulation of *CDK6* [168]. Likewise members of the *let-7* family, which are also often down-regulated in lung tumours, are involved in G_1 to S progression. This family consists of 12 members (in humans *let-7a* to *let-7k* and *miR-98*), which are encoded by 8 different loci and are almost identical in sequence. Currently the relative contribution of the different *let-7* family members to tumourigenesis is not entirely clear, as they have not been analysed in a combinatorial mode.

As already mentioned in the introduction, the complexity of the G_1 to S regulatory mechanism is extended by the fact that some miRNAs form part of regulatory loops. For example, both *let-7* and *miR-34* are able to down-regulate Myc [48, 80], while the latter protein blocks transcription of either one of these miRNAs [170]. Interestingly, Myc also seems to be involved in the transcriptional repression of *dLeu2*, the host gene of *miR-15a/16* [169]. On the other hand, Myc is able to induce the expression of *miR-17-92*. Moreover, Myc directly enhances the expression of CCND1, which in turn controls the expression of *miR-17-92* in a regulatory feedback loop [124]. However, CCND1 is also down-regulated by *miR-34a* and *miR-15a/16*. Likewise E2F, which is able to induce the transcription of *miR-17-92*, is regulated by *miR-34a* and *miR-15a/16*. All these regulatory loops are linked, since Myc is able to induce or repress all the miRNAs mentioned above as well as the transcription of *CCND1*. In accordance with these findings, a recent report even suggests that upregulation of Myc, which frequently occurs in cancer, could directly reprogram the miRNA transcriptome [170]. In addition, *miR-34a* family members and the miR-17-92 cluster are direct transcriptional targets of p53 [94, 171], the latter being often inactivated in tumours.

It is now becoming clear that such complex regulatory networks between miRNAs and their gene targets are actually common mechanisms that have evolved in mammals to enhance the robustness of gene regulation [172]. Thus miRNAs may not act primarily by reducing the expression of a few cancer regulatory genes, but rather by influencing the properties of the entire network in which these regulators may play a central role. As a consequence this may lead to the coordinated regulation of many genes. All these findings implicate that miRNAs may mainly act as stabilizers of biological networks by regulating hundreds of genes simultaneously. Currently, most effort has been made to elucidate the function and mechanism of regulation of individual miRNAs. However, since tumours usually deregulate many different miRNAs, we need to understand how microRNAs function in concert. Thus, to fully realize the potential of miRNAs, the combination of multiple miRNAs that target the same pathway should be considered for assessing a biological response rather than individual miRNAs.

Unmistakingly, miRNAs have a role in cancer and we need to recognize the possibility that their major impact is on networks rather than on individual targets. Further studies of the combinatorial expression and function of different miRNAs and their isoforms that are involved in similar processes will undoubtedly enhance the knowledge of how miRNAs act in concert to regulate cell cycle progression, cell death and other essential functions. Consequently, such studies should facilitate the identification of important key players involved in specific networks.
This comprehensive knowledge then should finally allow us to use miRNAs for future clinical applications like diagnosis, prognosis and potential therapy.

7. References

1. Jemal A, Thun MJ, Ries LA, Howe HL, Weir HK, Center MM, Ward E, Wu XC, Eheman C, Anderson R, Ajani UA, Kohler B, et al. Annual report to the nation on the status of cancer, 1975-2005, featuring trends in lung cancer, tobacco use, and tobacco control. J Natl Cancer Inst 2008;100:1672-94.
2. Travis WD, Travis LB, Devesa SS. Lung cancer. Cancer 1995;75:191-202.
3. Travis WD. Pathology of lung cancer. Clin Chest Med 2002;23:65-81, viii.
4. Fong KM, Sekido Y, Gazdar AF, Minna JD. Lung cancer. 9: Molecular biology of lung cancer: clinical implications. Thorax 2003;58:892-900.
5. Minna JD, Roth JA, Gazdar AF. Focus on lung cancer. Cancer Cell 2002;1:49-52.
6. Wu X, Piper-Hunter MG, Crawford M, Nuovo GJ, Marsh CB, Otterson GA, Nana-Sinkam SP. MicroRNAs in the pathogenesis of Lung Cancer. J Thorac Oncol 2009;4:1028-34.
7. Bunn PA, Jr., Shepherd FA, Sandler A, Le Chevalier T, Belani CP, Kosmidis PA, Scagliotti GV, Giaccone G. Ongoing and future trials of biologic therapies in lung cancer. Lung Cancer 2003;41 Suppl 1:S175-86.
8. Bilfinger TV. Surgical viewpoints for the definitive treatment of lung cancer. Respir Care Clin N Am 2003;9:141-62.
9. Socinski MA, Morris DE, Masters GA, Lilenbaum R. Chemotherapeutic management of stage IV non-small cell lung cancer. Chest 2003;123:226S-43S.
10. Schiller JH, Harrington D, Belani CP, Langer C, Sandler A, Krook J, Zhu J, Johnson DH. Comparison of four chemotherapy regimens for advanced non-small-cell lung cancer. N Engl J Med 2002;346:92-8.
11. Lee RC, Feinbaum RL, Ambros V. The C. elegans heterochronic gene lin-4 encodes small RNAs with antisense complementarity to lin-14. Cell 1993;75:843-54.
12. Pasquinelli AE, Reinhart BJ, Slack F, Martindale MQ, Kuroda MI, Maller B, Hayward DC, Ball EE, Degnan B, Muller P, Spring J, Srinivasan A, et al. Conservation of the sequence and temporal expression of let-7 heterochronic regulatory RNA. Nature 2000;408:86-9.
13. Lagos-Quintana M, Rauhut R, Lendeckel W, Tuschl T. Identification of novel genes coding for small expressed RNAs. Science 2001;294:853-8.
14. Lewis BP, Burge CB, Bartel DP. Conserved seed pairing, often flanked by adenosines, indicates that thousands of human genes are microRNA targets. Cell 2005;120:15-20.
15. Laurent LC. MicroRNAs in embryonic stem cells and early embryonic development. J Cell Mol Med 2008;12:2181-8.
16. Bueno MJ, Perez de Castro I, Malumbres M. Control of cell proliferation pathways by microRNAs. Cell Cycle 2008;7:3143-8.
17. Subramanian S, Steer CJ. MicroRNAs as gatekeepers of apoptosis. J Cell Physiol 2010;223:289-98.
18. Xu P, Vernooy SY, Guo M, Hay BA. The Drosophila microRNA Mir-14 suppresses cell death and is required for normal fat metabolism. Curr Biol 2003;13:790-5.
19. Tsitsiou E, Lindsay MA. microRNAs and the immune response. Curr Opin Pharmacol 2009;9:514-20.
20. Umbach JL, Cullen BR. The role of RNAi and microRNAs in animal virus replication and antiviral immunity. Genes Dev 2009;23:1151-64.
21. Navarro F, Lieberman J. Small RNAs guide hematopoietic cell differentiation and function. J Immunol 2010;184:5939-47.
22. Grosshans H, Slack FJ. Micro-RNAs: small is plentiful. J Cell Biol 2002;156:17-21.

23. Lim LP, Glasner ME, Yekta S, Burge CB, Bartel DP. Vertebrate microRNA genes. Science 2003;299:1540.

24. Lee CT, Risom T, Strauss WM. Evolutionary conservation of microRNA regulatory circuits: an examination of microRNA gene complexity and conserved microRNA-target interactions through metazoan phylogeny. DNA Cell Biol 2007;26:209-18.

25. Smalheiser NR. EST analyses predict the existence of a population of chimeric microRNA precursor-mRNA transcripts expressed in normal human and mouse tissues. Genome Biol 2003;4:403.

26. Cai X, Hagedorn CH, Cullen BR. Human microRNAs are processed from capped, polyadenylated transcripts that can also function as mRNAs. RNA 2004;10:1957-66.

27. Borchert GM, Lanier W, Davidson BL. RNA polymerase III transcribes human microRNAs. Nat Struct Mol Biol 2006;13:1097-101.

28. Lee Y, Ahn C, Han J, Choi H, Kim J, Yim J, Lee J, Provost P, Radmark O, Kim S, Kim VN. The nuclear RNase III Drosha initiates microRNA processing. Nature 2003;425:415-9.

29. Berezikov E, Chung WJ, Willis J, Cuppen E, Lai EC. Mammalian mirtron genes. Mol Cell 2007;28:328-36.

30. Yi R, Qin Y, Macara IG, Cullen BR. Exportin-5 mediates the nuclear export of pre-microRNAs and short hairpin RNAs. Genes Dev 2003;17:3011-6.

31. Lee Y, Jeon K, Lee JT, Kim S, Kim VN. MicroRNA maturation: stepwise processing and subcellular localization. EMBO J 2002;21:4663-70.

32. Lee Y, Hur I, Park SY, Kim YK, Suh MR, Kim VN. The role of PACT in the RNA silencing pathway. EMBO J 2006;25:522-32.

33. Khvorova A, Reynolds A, Jayasena SD. Functional siRNAs and miRNAs exhibit strand bias. Cell 2003;115:209-16.

34. Ebhardt HA, Tsang HH, Dai DC, Liu Y, Bostan B, Fahlman RP. Meta-analysis of small RNA-sequencing errors reveals ubiquitous post-transcriptional RNA modifications. Nucleic Acids Res 2009;37:2461-70.

35. Rhoades MW, Reinhart BJ, Lim LP, Burge CB, Bartel B, Bartel DP. Prediction of plant microRNA targets. Cell 2002;110:513-20.

36. Hutvagner G, Zamore PD. A microRNA in a multiple-turnover RNAi enzyme complex. Science 2002;297:2056-60.

37. Zeng Y, Yi R, Cullen BR. MicroRNAs and small interfering RNAs can inhibit mRNA expression by similar mechanisms. Proc Natl Acad Sci U S A 2003;100:9779-84.

38. Tay Y, Zhang J, Thomson AM, Lim B, Rigoutsos I. MicroRNAs to Nanog, Oct4 and Sox2 coding regions modulate embryonic stem cell differentiation. Nature 2008;455:1124-8.

39. Carthew RW, Sontheimer EJ. Origins and Mechanisms of miRNAs and siRNAs. Cell 2009;136:642-55.

40. Gu S, Kay MA. How do miRNAs mediate translational repression? Silence 2010;1:11.

41. Guo H, Ingolia NT, Weissman JS, Bartel DP. Mammalian microRNAs predominantly act to decrease target mRNA levels. Nature 2010;466:835-40.

42. Vasudevan S, Tong Y, Steitz JA. Switching from repression to activation: microRNAs can up-regulate translation. Science 2007;318:1931-4.

43. Calin GA, Sevignani C, Dumitru CD, Hyslop T, Noch E, Yendamuri S, Shimizu M, Rattan S, Bullrich F, Negrini M, Croce CM. Human microRNA genes are frequently located at fragile sites and genomic regions involved in cancers. Proc Natl Acad Sci U S A 2004;101:2999-3004.

44. Georges SA, Biery MC, Kim SY, Schelter JM, Guo J, Chang AN, Jackson AL, Carleton MO, Linsley PS, Cleary MA, Chau BN. Coordinated regulation of cell cycle transcripts by p53-Inducible microRNAs, miR-192 and miR-215. Cancer Res 2008;68:10105-12.

45. Aguda BD, Kim Y, Piper-Hunter MG, Friedman A, Marsh CB. MicroRNA regulation of a cancer network: consequences of the feedback loops

References

involving miR-17-92, E2F, and Myc. Proc Natl Acad Sci U S A 2008;105:19678-83.

46. Fabbri M. MicroRNAs and cancer epigenetics. Curr Opin Investig Drugs 2008;9:583-90.

47. Yu J, Wang F, Yang GH, Wang FL, Ma YN, Du ZW, Zhang JW. Human microRNA clusters: genomic organization and expression profile in leukemia cell lines. Biochem Biophys Res Commun 2006;349:59-68.

48. O'Donnell KA, Wentzel EA, Zeller KI, Dang CV, Mendell JT. c-Myc-regulated microRNAs modulate E2F1 expression. Nature 2005;435:839-43.

49. Sylvestre Y, De Guire V, Querido E, Mukhopadhyay UK, Bourdeau V, Major F, Ferbeyre G, Chartrand P. An E2F/miR-20a autoregulatory feedback loop. J Biol Chem 2007;282:2135-43.

50. Obernosterer G, Leuschner PJ, Alenius M, Martinez J. Post-transcriptional regulation of microRNA expression. RNA 2006;12:1161-7.

51. Wulczyn FG, Smirnova L, Rybak A, Brandt C, Kwidzinski E, Ninnemann O, Strehle M, Seiler A, Schumacher S, Nitsch R. Post-transcriptional regulation of the let-7 microRNA during neural cell specification. FASEB J 2007;21:415-26.

52. Santarpia L, Nicoloso M, Calin GA. MicroRNAs: a complex regulatory network drives the acquisition of malignant cell phenotype. Endocr Relat Cancer 2010;17:F51-75.

53. Lewis BP, Shih IH, Jones-Rhoades MW, Bartel DP, Burge CB. Prediction of mammalian microRNA targets. Cell 2003;115:787-98.

54. Hammell M, Long D, Zhang L, Lee A, Carmack CS, Han M, Ding Y, Ambros V. mirWIP: microRNA target prediction based on microRNA-containing ribonucleoprotein-enriched transcripts. Nat Methods 2008;5:813-9.

55. Barbarotto E, Schmittgen TD, Calin GA. MicroRNAs and cancer: profile, profile, profile. Int J Cancer 2008;122:969-77.

56. Hermeking H. p53 enters the microRNA world. Cancer Cell 2007;12:414-8.

57. Calin GA, Croce CM. MicroRNA signatures in human cancers. Nat Rev Cancer 2006;6:857-66.

58. Santarpia L, Nicoloso M, Calin GA. MicroRNAs: a complex regulatory network drives the acquisition of malignant cell phenotype. Endocr Relat Cancer;17:F51-75.

59. Hanahan D, Weinberg RA. The hallmarks of cancer. Cell 2000;100:57-70.

60. Izzotti A, Calin GA, Steele VE, Croce CM, De Flora S. Relationships of microRNA expression in mouse lung with age and exposure to cigarette smoke and light. FASEB J 2009;23:3243-50.

61. Schembri F, Sridhar S, Perdomo C, Gustafson AM, Zhang X, Ergun A, Lu J, Liu G, Bowers J, Vaziri C, Ott K, Sensinger K, et al. MicroRNAs as modulators of smoking-induced gene expression changes in human airway epithelium. Proc Natl Acad Sci U S A 2009;106:2319-24.

62. Bishop JA, Benjamin H, Cholakh H, Chajut A, Clark DP, Westra WH. Accurate classification of non-small cell lung carcinoma using a novel microRNA-based approach. Clin Cancer Res 2010;16:610-9.

63. Lebanony D, Benjamin H, Gilad S, Ezagouri M, Dov A, Ashkenazi K, Gefen N, Izraeli S, Rechavi G, Pass H, Nonaka D, Li J, et al. Diagnostic assay based on hsa-miR-205 expression distinguishes squamous from nonsquamous non-small-cell lung carcinoma. J Clin Oncol 2009;27:2030-7.

64. Miko E, Czimmerer Z, Csanky E, Boros G, Buslig J, Dezso B, Scholtz B. Differentially expressed microRNAs in small cell lung cancer. Exp Lung Res 2009;35:646-64.

65. Yanaihara N, Caplen N, Bowman E, Seike M, Kumamoto K, Yi M, Stephens RM, Okamoto A, Yokota J, Tanaka T, Calin GA, Liu CG, et al. Unique microRNA molecular profiles in lung cancer diagnosis and prognosis. Cancer Cell 2006;9:189-98.

66. Volinia S, Calin GA, Liu CG, Ambs S, Cimmino A, Petrocca F, Visone R, Iorio M, Roldo C, Ferracin M, Prueitt RL,

Yanaihara N, et al. A microRNA expression signature of human solid tumors defines cancer gene targets. Proc Natl Acad Sci U S A 2006;103:2257-61.

67. Navarro A, Marrades RM, Vinolas N, Quera A, Agusti C, Huerta A, Ramirez J, Torres A, Monzo M. MicroRNAs expressed during lung cancer development are expressed in human pseudoglandular lung embryogenesis. Oncology 2009;76:162-9.

68. Seike M, Goto A, Okano T, Bowman ED, Schetter AJ, Horikawa I, Mathe EA, Jen J, Yang P, Sugimura H, Gemma A, Kudoh S, et al. MiR-21 is an EGFR-regulated anti-apoptotic factor in lung cancer in never-smokers. Proc Natl Acad Sci U S A 2009;106:12085-90.

69. Mascaux C, Laes JF, Anthoine G, Haller A, Ninane V, Burny A, Sculier JP. Evolution of microRNA expression during human bronchial squamous carcinogenesis. Eur Respir J 2009;33:352-9.

70. Gao W, Yu Y, Cao H, Shen H, Li X, Pan S, Shu Y. Deregulated expression of miR-21, miR-143 and miR-181a in non small cell lung cancer is related to clinicopathologic characteristics or patient prognosis. Biomed Pharmacother 2010;64:399-408.

71. Yu SL, Chen HY, Chang GC, Chen CY, Chen HW, Singh S, Cheng CL, Yu CJ, Lee YC, Chen HS, Su TJ, Chiang CC, et al. MicroRNA signature predicts survival and relapse in lung cancer. Cancer Cell 2008;13:48-57.

72. Nasser MW, Datta J, Nuovo G, Kutay H, Motiwala T, Majumder S, Wang B, Suster S, Jacob ST, Ghoshal K. Down-regulation of micro-RNA-1 (miR-1) in lung cancer. Suppression of tumorigenic property of lung cancer cells and their sensitization to doxorubicin-induced apoptosis by miR-1. J Biol Chem 2008;283:33394-405.

73. Takamizawa J, Konishi H, Yanagisawa K, Tomida S, Osada H, Endoh H, Harano T, Yatabe Y, Nagino M, Nimura Y, Mitsudomi T, Takahashi T. Reduced expression of the let-7 microRNAs in human lung cancers in association with shortened postoperative survival. Cancer Res 2004;64:3753-6.

74. Chin LJ, Ratner E, Leng S, Zhai R, Nallur S, Babar I, Muller RU, Straka E, Su L, Burki EA, Crowell RE, Patel R, et al. A SNP in a let-7 microRNA complementary site in the KRAS 3' untranslated region increases non-small cell lung cancer risk. Cancer Res 2008;68:8535-40.

75. Xie Y, Todd NW, Liu Z, Zhan M, Fang H, Peng H, Alattar M, Deepak J, Stass SA, Jiang F. Altered miRNA expression in sputum for diagnosis of non-small cell lung cancer. Lung Cancer 2010;67:170-6.

76. Hu Z, Chen X, Zhao Y, Tian T, Jin G, Shu Y, Chen Y, Xu L, Zen K, Zhang C, Shen H. Serum microRNA signatures identified in a genome-wide serum microRNA expression profiling predict survival of non-small-cell lung cancer. J Clin Oncol 2010;28:1721-6.

77. Ortholan C, Puissegur MP, Ilie M, Barbry P, Mari B, Hofman P. MicroRNAs and lung cancer: new oncogenes and tumor suppressors, new prognostic factors and potential therapeutic targets. Curr Med Chem 2009;16:1047-61.

78. Tang Y, Zheng J, Sun Y, Wu Z, Liu Z, Huang G. MicroRNA-1 regulates cardiomyocyte apoptosis by targeting Bcl-2. Int Heart J 2009;50:377-87.

79. Johnson SM, Grosshans H, Shingara J, Byrom M, Jarvis R, Cheng A, Labourier E, Reinert KL, Brown D, Slack FJ. RAS is regulated by the let-7 microRNA family. Cell 2005;120:635-47.

80. Sampson VB, Rong NH, Han J, Yang Q, Aris V, Soteropoulos P, Petrelli NJ, Dunn SP, Krueger LJ. MicroRNA let-7a down-regulates MYC and reverts MYC-induced growth in Burkitt lymphoma cells. Cancer Res 2007;67:9762-70.

81. Lee YS, Dutta A. The tumor suppressor microRNA let-7 represses the HMGA2 oncogene. Genes Dev 2007;21:1025-30.

82. Du L, Schageman JJ, Subauste MC, Saber B, Hammond SM, Prudkin L, Wistuba, II, Ji L, Roth JA, Minna JD, Pertsemlidis A. miR-93, miR-98, and miR-197 regulate expression of tumor suppressor gene FUS1. Mol Cancer Res 2009;7:1234-43.

References

83. Bandi N, Zbinden S, Gugger M, Arnold M, Kocher V, Hasan L, Kappeler A, Brunner T, Vassella E. miR-15a and miR-16 are implicated in cell cycle regulation in a Rb-dependent manner and are frequently deleted or down-regulated in non-small cell lung cancer. Cancer Res 2009;69:5553-9.

84. Bonci D, Coppola V, Musumeci M, Addario A, Giuffrida R, Memeo L, D'Urso L, Pagliuca A, Biffoni M, Labbaye C, Bartucci M, Muto G, et al. The miR-15a-miR-16-1 cluster controls prostate cancer by targeting multiple oncogenic activities. Nat Med 2008;14:1271-7.

85. Liu Q, Fu H, Sun F, Zhang H, Tie Y, Zhu J, Xing R, Sun Z, Zheng X. miR-16 family induces cell cycle arrest by regulating multiple cell cycle genes. Nucleic Acids Res 2008;36:5391-404.

86. Cimmino A, Calin GA, Fabbri M, Iorio MV, Ferracin M, Shimizu M, Wojcik SE, Aqeilan RI, Zupo S, Dono M, Rassenti L, Alder H, et al. miR-15 and miR-16 induce apoptosis by targeting BCL2. Proc Natl Acad Sci U S A 2005;102:13944-9.

87. Lee SO, Masyuk T, Splinter P, Banales JM, Masyuk A, Stroope A, Larusso N. MicroRNA15a modulates expression of the cell-cycle regulator Cdc25A and affects hepatic cystogenesis in a rat model of polycystic kidney disease. J Clin Invest 2008;118:3714-24.

88. Bhattacharya R, Nicoloso M, Arvizo R, Wang E, Cortez A, Rossi S, Calin GA, Mukherjee P. MiR-15a and MiR-16 control Bmi-1 expression in ovarian cancer. Cancer Res 2009;69:9090-5.

89. Fabbri M, Garzon R, Cimmino A, Liu Z, Zanesi N, Callegari E, Liu S, Alder H, Costinean S, Fernandez-Cymering C, Volinia S, Guler G, et al. MicroRNA-29 family reverts aberrant methylation in lung cancer by targeting DNA methyltransferases 3A and 3B. Proc Natl Acad Sci U S A 2007;104:15805-10.

90. Mott JL, Kobayashi S, Bronk SF, Gores GJ. mir-29 regulates Mcl-1 protein expression and apoptosis. Oncogene 2007;26:6133-40.

91. Gallardo E, Navarro A, Vinolas N, Marrades RM, Diaz T, Gel B, Quera A, Bandres E, Garcia-Foncillas J, Ramirez J, Monzo M. miR-34a as a prognostic marker of relapse in surgically resected non-small-cell lung cancer. Carcinogenesis 2009;30:1903-9.

92. Sun F, Fu H, Liu Q, Tie Y, Zhu J, Xing R, Sun Z, Zheng X. Downregulation of CCND1 and CDK6 by miR-34a induces cell cycle arrest. FEBS Lett 2008;582:1564-8.

93. He L, He X, Lim LP, de Stanchina E, Xuan Z, Liang Y, Xue W, Zender L, Magnus J, Ridzon D, Jackson AL, Linsley PS, et al. A microRNA component of the p53 tumour suppressor network. Nature 2007;447:1130-4.

94. Bommer GT, Gerin I, Feng Y, Kaczorowski AJ, Kuick R, Love RE, Zhai Y, Giordano TJ, Qin ZS, Moore BB, MacDougald OA, Cho KR, et al. p53-mediated activation of miRNA34 candidate tumor-suppressor genes. Curr Biol 2007;17:1298-307.

95. Cannell IG, Kong YW, Johnston SJ, Chen ML, Collins HM, Dobbyn HC, Elia A, Kress TR, Dickens M, Clemens MJ, Heery DM, Gaestel M, et al. p38 MAPK/MK2-mediated induction of miR-34c following DNA damage prevents Myc-dependent DNA replication. Proc Natl Acad Sci U S A 2010;107:5375-80.

96. Welch C, Chen Y, Stallings RL. MicroRNA-34a functions as a potential tumor suppressor by inducing apoptosis in neuroblastoma cells. Oncogene 2007;26:5017-22.

97. Yamakuchi M, Lowenstein CJ. MiR-34, SIRT1 and p53: the feedback loop. Cell Cycle 2009;8:712-5.

98. Crawford M, Brawner E, Batte K, Yu L, Hunter MG, Otterson GA, Nuovo G, Marsh CB, Nana-Sinkam SP. MicroRNA-126 inhibits invasion in non-small cell lung carcinoma cell lines. Biochem Biophys Res Commun 2008;373:607-12.

99. Liu B, Peng XC, Zheng XL, Wang J, Qin YW. MiR-126 restoration down-regulate VEGF and inhibit the growth of lung

cancer cell lines in vitro and in vivo. Lung Cancer 2009;66:169-75.

100. Weiss GJ, Bemis LT, Nakajima E, Sugita M, Birks DK, Robinson WA, Varella-Garcia M, Bunn PA, Jr., Haney J, Helfrich BA, Kato H, Hirsch FR, et al. EGFR regulation by microRNA in lung cancer: correlation with clinical response and survival to gefitinib and EGFR expression in cell lines. Ann Oncol 2008;19:1053-9.

101. Wang G, Mao W, Zheng S. MicroRNA-183 regulates Ezrin expression in lung cancer cells. FEBS Lett 2008;582:3663-8.

102. Ceppi P, Mudduluru G, Kumarswamy R, Rapa I, Scagliotti GV, Papotti M, Allgayer H. Loss of miR-200c Expression Induces an Aggressive, Invasive, and Chemoresistant Phenotype in Non-Small Cell Lung Cancer. Mol Cancer Res 2010;8:1207-16.

103. Korpal M, Lee ES, Hu G, Kang Y. The miR-200 family inhibits epithelial-mesenchymal transition and cancer cell migration by direct targeting of E-cadherin transcriptional repressors ZEB1 and ZEB2. J Biol Chem 2008;283:14910-4.

104. Zhang JG, Wang JJ, Zhao F, Liu Q, Jiang K, Yang GH. MicroRNA-21 (miR-21) represses tumor suppressor PTEN and promotes growth and invasion in non-small cell lung cancer (NSCLC). Clin Chim Acta 2010;411:846-52.

105. Talotta F, Cimmino A, Matarazzo MR, Casalino L, De Vita G, D'Esposito M, Di Lauro R, Verde P. An autoregulatory loop mediated by miR-21 and PDCD4 controls the AP-1 activity in RAS transformation. Oncogene 2009;28:73-84.

106. Zhu S, Si ML, Wu H, Mo YY. MicroRNA-21 targets the tumor suppressor gene tropomyosin 1 (TPM1). J Biol Chem 2007;282:14328-36.

107. Hayashita Y, Osada H, Tatematsu Y, Yamada H, Yanagisawa K, Tomida S, Yatabe Y, Kawahara K, Sekido Y, Takahashi T. A polycistronic microRNA cluster, miR-17-92, is overexpressed in human lung cancers and enhances cell proliferation. Cancer Res 2005;65:9628-32.

108. Xiao C, Srinivasan L, Calado DP, Patterson HC, Zhang B, Wang J, Henderson JM, Kutok JL, Rajewsky K. Lymphoproliferative disease and autoimmunity in mice with increased miR-17-92 expression in lymphocytes. Nat Immunol 2008;9:405-14.

109. Rodriguez A, Vigorito E, Clare S, Warren MV, Couttet P, Soond DR, van Dongen S, Grocock RJ, Das PP, Miska EA, Vetrie D, Okkenhaug K, et al. Requirement of bic/microRNA-155 for normal immune function. Science 2007;316:608-11.

110. Bolisetty MT, Dy G, Tam W, Beemon KL. Reticuloendotheliosis virus strain T induces miR-155, which targets JARID2 and promotes cell survival. J Virol 2009;83:12009-17.

111. Garofalo M, Di Leva G, Romano G, Nuovo G, Suh SS, Ngankeu A, Taccioli C, Pichiorri F, Alder H, Secchiero P, Gasparini P, Gonelli A, et al. miR-221&222 regulate TRAIL resistance and enhance tumorigenicity through PTEN and TIMP3 downregulation. Cancer Cell 2009;16:498-509.

112. Frenquelli M, Muzio M, Scielzo C, Fazi C, Scarfo L, Rossi C, Ferrari G, Ghia P, Caligaris-Cappio F. MicroRNA and proliferation control in chronic lymphocytic leukemia: functional relationship between miR-221/222 cluster and p27. Blood 2010;115:3949-59.

113. Kim H, Kwon YM, Kim JS, Han J, Shim YM, Park J, Kim DH. Elevated mRNA levels of DNA methyltransferase-1 as an independent prognostic factor in primary nonsmall cell lung cancer. Cancer 2006;107:1042-9.

114. Esteller M. Epigenetics in cancer. N Engl J Med 2008;358:1148-59.

115. Johnson CD, Esquela-Kerscher A, Stefani G, Byrom M, Kelnar K, Ovcharenko D, Wilson M, Wang X, Shelton J, Shingara J, Chin L, Brown D, et al. The let-7 microRNA represses cell proliferation pathways in human cells. Cancer Res 2007;67:7713-22.

116. Kumar MS, Erkeland SJ, Pester RE, Chen CY, Ebert MS, Sharp PA, Jacks T. Suppression of non-small cell lung tumor development by the let-7 microRNA

References

family. Proc Natl Acad Sci U S A 2008;105:3903-8.

117. Cho WC, Chow AS, Au JS. Restoration of tumour suppressor hsa-miR-145 inhibits cancer cell growth in lung adenocarcinoma patients with epidermal growth factor receptor mutation. Eur J Cancer 2009;45:2197-206.

118. Wiggins JF, Ruffino L, Kelnar K, Omotola M, Patrawala L, Brown D, Bader AG. Development of a lung cancer therapeutic based on the tumor suppressor microRNA-34. Cancer Res 2010;70:5923-30.

119. Esquela-Kerscher A, Trang P, Wiggins JF, Patrawala L, Cheng A, Ford L, Weidhaas JB, Brown D, Bader AG, Slack FJ. The let-7 microRNA reduces tumor growth in mouse models of lung cancer. Cell Cycle 2008;7:759-64.

120. Osada H, Takahashi T. Genetic alterations of multiple tumor suppressors and oncogenes in the carcinogenesis and progression of lung cancer. Oncogene 2002;21:7421-34.

121. Lujambio A, Esteller M. CpG island hypermethylation of tumor suppressor microRNAs in human cancer. Cell Cycle 2007;6:1455-9.

122. Karube Y, Tanaka H, Osada H, Tomida S, Tatematsu Y, Yanagisawa K, Yatabe Y, Takamizawa J, Miyoshi S, Mitsudomi T, Takahashi T. Reduced expression of Dicer associated with poor prognosis in lung cancer patients. Cancer Sci 2005;96:111-5.

123. Chen D, Farwell MA, Zhang B. MicroRNA as a new player in the cell cycle. J Cell Physiol 2010;225:296-301.

124. Yu Z, Wang C, Wang M, Li Z, Casimiro MC, Liu M, Wu K, Whittle J, Ju X, Hyslop T, McCue P, Pestell RG. A cyclin D1/microRNA 17/20 regulatory feedback loop in control of breast cancer cell proliferation. J Cell Biol 2008;182:509-17.

125. Viswanathan SR, Daley GQ, Gregory RI. Selective blockade of microRNA processing by Lin28. Science 2008;320:97-100.

126. Hwang HW, Wentzel EA, Mendell JT. A hexanucleotide element directs microRNA nuclear import. Science 2007;315:97-100.

127. Sandberg R, Neilson JR, Sarma A, Sharp PA, Burge CB. Proliferating cells express mRNAs with shortened 3' untranslated regions and fewer microRNA target sites. Science 2008;320:1643-7.

128. Kedde M, Agami R. Interplay between microRNAs and RNA-binding proteins determines developmental processes. Cell Cycle 2008;7:899-903.

129. Eulalio A, Tritschler F, Izaurralde E. The GW182 protein family in animal cells: new insights into domains required for miRNA-mediated gene silencing. RNA 2009;15:1433-42.

130. Davis BN, Hilyard AC, Lagna G, Hata A. SMAD proteins control DROSHA-mediated microRNA maturation. Nature 2008;454:56-61.

131. Calin GA, Dumitru CD, Shimizu M, Bichi R, Zupo S, Noch E, Aldler H, Rattan S, Keating M, Rai K, Rassenti L, Kipps T, et al. Frequent deletions and down-regulation of micro- RNA genes miR15 and miR16 at 13q14 in chronic lymphocytic leukemia. Proc Natl Acad Sci U S A 2002;99:15524-9.

132. Stilgenbauer S, Nickolenko J, Wilhelm J, Wolf S, Weitz S, Dohner K, Boehm T, Dohner H, Lichter P. Expressed sequences as candidates for a novel tumor suppressor gene at band 13q14 in B-cell chronic lymphocytic leukemia and mantle cell lymphoma. Oncogene 1998;16:1891-7.

133. Corthals SL, Jongen-Lavrencic M, de Knegt Y, Peeters JK, Beverloo HB, Lokhorst HM, Sonneveld P. Micro-RNA-15a and micro-RNA-16 expression and chromosome 13 deletions in multiple myeloma. Leuk Res 2010;34:677-81.

134. Chen C, Frierson HF, Jr., Haggerty PF, Theodorescu D, Gregory CW, Dong JT. An 800-kb region of deletion at 13q14 in human prostate and other carcinomas. Genomics 2001;77:135-44.

135. Bottoni A, Piccin D, Tagliati F, Luchin A, Zatelli MC, degli Uberti EC. miR-15a and

miR-16-1 down-regulation in pituitary adenomas. J Cell Physiol 2005;204:280-5.

136. Calin GA, Cimmino A, Fabbri M, Ferracin M, Wojcik SE, Shimizu M, Taccioli C, Zanesi N, Garzon R, Aqeilan RI, Alder H, Volinia S, et al. MiR-15a and miR-16-1 cluster functions in human leukemia. Proc Natl Acad Sci U S A 2008;105:5166-71.

137. Linsley PS, Schelter J, Burchard J, Kibukawa M, Martin MM, Bartz SR, Johnson JM, Cummins JM, Raymond CK, Dai H, Chau N, Cleary M, et al. Transcripts targeted by the microRNA-16 family cooperatively regulate cell cycle progression. Mol Cell Biol 2007;27:2240-52.

138. Chen RW, Bemis LT, Amato CM, Myint H, Tran H, Birks DK, Eckhardt SG, Robinson WA. Truncation in CCND1 mRNA alters miR-16-1 regulation in mantle cell lymphoma. Blood 2008;112:822-9.

139. Lodygin D, Tarasov V, Epanchintsev A, Berking C, Knyazeva T, Korner H, Knyazev P, Diebold J, Hermeking H. Inactivation of miR-34a by aberrant CpG methylation in multiple types of cancer. Cell Cycle 2008;7:2591-600.

140. Fujita Y, Kojima K, Hamada N, Ohhashi R, Akao Y, Nozawa Y, Deguchi T, Ito M. Effects of miR-34a on cell growth and chemoresistance in prostate cancer PC3 cells. Biochem Biophys Res Commun 2008;377:114-9.

141. Tazawa H, Tsuchiya N, Izumiya M, Nakagama H. Tumor-suppressive miR-34a induces senescence-like growth arrest through modulation of the E2F pathway in human colon cancer cells. Proc Natl Acad Sci U S A 2007;104:15472-7.

142. Zenz T, Mohr J, Eldering E, Kater AP, Buhler A, Kienle D, Winkler D, Durig J, van Oers MH, Mertens D, Dohner H, Stilgenbauer S. miR-34a as part of the resistance network in chronic lymphocytic leukemia. Blood 2009;113:3801-8.

143. Asslaber D, Pinon JD, Seyfried I, Desch P, Stocher M, Tinhofer I, Egle A, Merkel O, Greil R. microRNA-34a expression correlates with MDM2 SNP309 polymorphism and treatment-free survival in chronic lymphocytic leukemia. Blood 2010;115:4191-7.

144. Wei JS, Song YK, Durinck S, Chen QR, Cheuk AT, Tsang P, Zhang Q, Thiele CJ, Slack A, Shohet J, Khan J. The MYCN oncogene is a direct target of miR-34a. Oncogene 2008;27:5204-13.

145. Yamakuchi M, Ferlito M, Lowenstein CJ. miR-34a repression of SIRT1 regulates apoptosis. Proc Natl Acad Sci U S A 2008;105:13421-6.

146. Calin GA, Ferracin M, Cimmino A, Di Leva G, Shimizu M, Wojcik SE, Iorio MV, Visone R, Sever NI, Fabbri M, Iuliano R, Palumbo T, et al. A MicroRNA signature associated with prognosis and progression in chronic lymphocytic leukemia. N Engl J Med 2005;353:1793-801.

147. Shen J, Ambrosone CB, DiCioccio RA, Odunsi K, Lele SB, Zhao H. A functional polymorphism in the miR-146a gene and age of familial breast/ovarian cancer diagnosis. Carcinogenesis 2008;29:1963-6.

148. Yu Z, Li Z, Jolicoeur N, Zhang L, Fortin Y, Wang E, Wu M, Shen SH. Aberrant allele frequencies of the SNPs located in microRNA target sites are potentially associated with human cancers. Nucleic Acids Res 2007;35:4535-41.

149. Negrini M, Ferracin M, Sabbioni S, Croce CM. MicroRNAs in human cancer: from research to therapy. J Cell Sci 2007;120:1833-40.

150. Heneghan HM, Miller N, Kerin MJ. MiRNAs as biomarkers and therapeutic targets in cancer. Curr Opin Pharmacol 2010;[Epub ahead of print].

151. Zhang S, Chen L, Jung EJ, Calin GA. Targeting microRNAs with small molecules: from dream to reality. Clin Pharmacol Ther 2010;87:754-8.

152. Sarkar FH, Li Y, Wang Z, Kong D, Ali S. Implication of microRNAs in drug resistance for designing novel cancer therapy. Drug Resist Updat 2010;13:57-66.

153. Seto AG. The road toward microRNA therapeutics. Int J Biochem Cell Biol 2010;42:1298-305.

154. Petri A, Lindow M, Kauppinen S. MicroRNA silencing in primates: towards

References

development of novel therapeutics. Cancer Res 2009;69:393-5.

155. Hummel R, Hussey DJ, Haier J. MicroRNAs: predictors and modifiers of chemo- and radiotherapy in different tumour types. Eur J Cancer 2010;46:298-311.

156. Kota J, Chivukula RR, O'Donnell KA, Wentzel EA, Montgomery CL, Hwang HW, Chang TC, Vivekanandan P, Torbenson M, Clark KR, Mendell JR, Mendell JT. Therapeutic microRNA delivery suppresses tumorigenesis in a murine liver cancer model. Cell 2009;137:1005-17.

157. Wang X, Tang S, Le SY, Lu R, Rader JS, Meyers C, Zheng ZM. Aberrant expression of oncogenic and tumor-suppressive microRNAs in cervical cancer is required for cancer cell growth. PLoS One 2008;3:e2557.

158. Zhang H, Luo XQ, Zhang P, Huang LB, Zheng YS, Wu J, Zhou H, Qu LH, Xu L, Chen YQ. MicroRNA patterns associated with clinical prognostic parameters and CNS relapse prediction in pediatric acute leukemia. PLoS One 2009;4:e7826.

159. Robbs BK, Cruz AL, Werneck MB, Mognol GP, Viola JP. Dual roles for NFAT transcription factor genes as oncogenes and tumor suppressors. Mol Cell Biol 2008;28:7168-81.

160. Lane DP, Benchimol S. p53: oncogene or anti-oncogene? Genes Dev 1990;4:1-8.

161. Johnson DG. The paradox of E2F1: oncogene and tumor suppressor gene. Mol Carcinog 2000;27:151-7.

162. Hermeking H. MiR-34a and p53. Cell Cycle 2009;8:1308.

163. Suzuki HI, Yamagata K, Sugimoto K, Iwamoto T, Kato S, Miyazono K. Modulation of microRNA processing by p53. Nature 2009;460:529-33.

164. Tammemagi MC, McLaughlin JR, Bull SB. Meta-analyses of p53 tumor suppressor gene alterations and clinicopathological features in resected lung cancers. Cancer Epidemiol Biomarkers Prev 1999;8:625-34.

165. Leversha MA, Fielding P, Watson S, Gosney JR, Field JK. Expression of p53, pRB, and p16 in lung tumours: a validation study on tissue microarrays. J Pathol 2003;200:610-9.

166. Morgan DO. Cyclin-dependent kinases: engines, clocks, and microprocessors. Annu Rev Cell Dev Biol 1997;13:261-91.

167. Baek D, Villen J, Shin C, Camargo FD, Gygi SP, Bartel DP. The impact of microRNAs on protein output. Nature 2008;455:64-71.

168. Lujambio A, Ropero S, Ballestar E, Fraga MF, Cerrato C, Setien F, Casado S, Suarez-Gauthier A, Sanchez-Cespedes M, Git A, Spiteri I, Das PP, et al. Genetic unmasking of an epigenetically silenced microRNA in human cancer cells. Cancer Res 2007;67:1424-9.

169. Lerner M, Harada M, Loven J, Castro J, Davis Z, Oscier D, Henriksson M, Sangfelt O, Grander D, Corcoran MM. DLEU2, frequently deleted in malignancy, functions as a critical host gene of the cell cycle inhibitory microRNAs miR-15a and miR-16-1. Exp Cell Res 2009;315:2941-52.

170. Chang TC, Yu D, Lee YS, Wentzel EA, Arking DE, West KM, Dang CV, Thomas-Tikhonenko A, Mendell JT. Widespread microRNA repression by Myc contributes to tumorigenesis. Nat Genet 2008;40:43-50.

171. Yan HL, Xue G, Mei Q, Wang YZ, Ding FX, Liu MF, Lu MH, Tang Y, Yu HY, Sun SH. Repression of the miR-17-92 cluster by p53 has an important function in hypoxia-induced apoptosis. EMBO J 2009;28:2719-32.

172. Tsang J, Zhu J, van Oudenaarden A. MicroRNA-mediated feedback and feedforward loops are recurrent network motifs in mammals. Mol Cell 2007;26:753-67.

8. Acknowledgments

..... to Erik Vassella for his guidance and support during the last years.

.... to Samuel Zbinden for his work for the *miR-15a/16* part of this thesis and many funny hours in the lab.

.... to all current and former master students, who helped the Vassella lab to be a pleasant place to work.

... to Verena Kocher, Claudia Zurbuchen and Karin Blaser for their help and support and for creating a warm atmosphere in the lab.

... to the Swiss National Foundation for supporting my work.

.... to Thomi Brunner for his optimistic view and his help and advice in technical and biological questions.

... to Nadia Corazza for constructive discussions about scientific and real life.

... to all the current and former members of the Brunner and the Müller lab for many interesting and motivating discussions and coffee breaks.

... to Sacha Kämpfer, Karin Stettler and Carmen Brand for interesting and encouraging lunch hours.

... to Pablo Hess, for his endless support and understanding that quality time has often been scarce.

... to my parents for being always there, whenever I needed them.

Die VDM Verlagsservicegesellschaft sucht für wissenschaftliche Verlage abgeschlossene und herausragende

Dissertationen, Habilitationen, Diplomarbeiten, Master Theses, Magisterarbeiten usw.

für die kostenlose Publikation als Fachbuch.

Sie verfügen über eine Arbeit, die hohen inhaltlichen und formalen Ansprüchen genügt, und haben Interesse an einer honorarvergüteten Publikation?

Dann senden Sie bitte erste Informationen über sich und Ihre Arbeit per Email an *info@vdm-vsg.de*.

Sie erhalten kurzfristig unser Feedback!

VDM Verlagsservicegesellschaft mbH
Dudweiler Landstr. 99   Telefon +49 681 3720 174
D - 66123 Saarbrücken   Fax   +49 681 3720 1749
www.vdm-vsg.de

Die VDM Verlagsservicegesellschaft mbH vertritt

Printed by Books on Demand GmbH, Norderstedt / Germany